MATHEMATICS OF ADAPTIVE CONTROL PROCESSES

MODERN ANALYTIC AND COMPUTATIONAL METHODS IN SCIENCE AND MATHEMATICS
A Group of Monographs and Advanced Textbooks

Editor: RICHARD BELLMAN, University of Southern California

MODERN ANALYTIC AND COMPUTATIONAL METHODS IN SCIENCE AND MATHEMATICS

MÉTHODES MODERNES D'ANALYSE ET DE COMPUTATION EN SCIENCE ET MATHÉMATIQUE

NEUE ANALYTISCHE UND NUMERISCHE METHODEN IN DER WISSENSCHAFT UND DER MATHMÈATIK

НОВЫЕ АНАЛИТИЧЕСКИЕ И БЫЧИСЛИТЕЛЬНЫЕ МЕТОДЫ В НАУКЕ И МАТЕМАТИКЕ

Editor

RICHARD BELLMAN, UNIVERSITY OF SOUTHERN CALIFORNIA

Mathematics of
Adaptive Control Processes

by

SIDNEY J. YAKOWITZ
The University of Arizona, Tucson, Arizona

American Elsevier Publishing Company, Inc.
New York   1969

AMERICAN ELSEVIER PUBLISHING COMPANY, INC.
52 Vanderbilt Avenue
New York, N.Y. 10017

ELSEVIER PUBLISHING COMPANY, LTD.
Barking, Essex, England

ELSEVIER PUBLISHING COMPANY
335 Jan Van Galenstraat, P.O. Box 211
Amsterdam, The Netherlands

Standard Book Number 444-00048-8
Library of Congress Card Number 68-26813

. . . To Rosemary
this book owes its being . . .

PREFACE

The primary objective of this book is to present a rigorous, unified, and inclusive systems theory for multi-stage decision processes. In this theory the principle of optimality occurs as a theorem and the dynamic programming algorithm is revealed in a context which makes plain its domain of proper application. Such a systems theory is developed in Chapter 2 for deterministic systems (control processes) and extended to probabilistic systems (adaptive control processes) in Chapter 3.

The purpose of the remainder of the book is twofold. First, I wish to show the generality of adaptive control process theory by demonstrating that decision processes of great research interest—two-armed bandits and supervised learning in statistical communication theory—are conveniently modeled as adaptive control processes. Second, I seek to impress the reader with the benefits of studying the logical structure of adaptive control processes by using the adaptive control process theory developed in Chapter 3 to make substantial contributions to the two-armed bandit problem (Chapter 4) and to the theory of supervised learning in statistical communications (Chapter 5). I contend that the research contributions made to these disciplines would have been worthy of attention even if the derivations had been conventional.

I have been compelled to undertake these studies after reflecting on similarities of three areas of scientific inquiry—information theory, sequential decision theory, and dynamic programming (or multi-stage decision) processes. Research in all these fields was initiated in the late 1940's, and each of these fields has been dominated by one or two outstanding investigators: Claude Shannon has made the principal statements about information theory; Abraham Wald and Herbert Robbins have been the eloquent spokesmen for sequential decision theory; and Richard Bellman's studies began and still dominate the dynamic programming literature. Although I have not read any papers strongly connecting any two (much less all three) of these disciplines, it occurred to me that structurally as well as historically they have much in common. The problems in each field are problems in optimization in a statistical context, and decisions are made sequentially rather than continuously (variational calculus) or once (game theory). Further, application of classic mathematics of optimization to these areas is awkward or sometimes totally inappropriate. Finally, all three disciplines are especially suited to (and would scarcely be expected to exist apart from) digital computers. As a consequence (in contrast to variational calculus and other classic methods of optimization), in information theory, sequential decision

theory, and dynamic programming one is not so much interested in uniqueness and characterization theorems as effective algorithms for obtaining
some result in finitely many iterations. It was these considerations which led
me to think that these fields have a common underlying mathematical
structure, and methods successful in one would be useful in the others. I
believe this text shows that my hunch was well founded.

There is reason to hope that students and researchers studying decision
processes in mathematical economics, engineering, or statistics will find the
careful development of adaptive control process theory presented here as
valuable for their studies as I have found it in my investigations in two-
armed bandit and pattern recognition problems. In addition, interested
mathematicians may find this book suitable as a concise introduction to
fundamental concepts of control theory, sequential analysis, and statistical
communication theory.

The exposition does not assume the reader is a specialist in any of the fields
of discussion, but supposes only a background in probability theory and a
mind that does not balk at abstractions.

I have used the chapters on control processes and adaptive control processes, in conjunction with *Applied Dynamic Programming* by Bellman and
Dreyfus, as text for a three-unit senior-graduate level course in optimization.
Exercises are included in this book with the intention of their being used
either for homework assignments or tests of mastery for the lone reader.
The exercises carry the burden of indicating the engineering significance of
the theory in the text, which I have taken great pains to make concise and
lean. Included also are a number of generic numerical examples worked out
in detail (and often presented with a computer program and readout) which
demonstrate the effectiveness of my methods.

This book is based on my Ph.D. dissertation (electrical engineering,
Arizona State University). It all began while I was browsing through the
library stacks and was struck by the odd title *Adaptive Control Processes,
A Guided Tour* (Bellman). I took the tour and returned on my own to inspect
two-armed bandits more carefully. When I had come to master the two-
armed bandit problem, after a fashion, I recognized that the supervised
learning problem in pattern recognition theory could be solved in full
generality by the same techniques. My thesis advisor was satisfied that this
was a dissertation's worth of discovery. In the process of writing this, to my
astonishment after a vigorous and prolonged literature search I failed to
find logical justification for the methods I had been using and which enjoy
considerable popularity in the engineering literature. This state of affairs
struck me as unsatisfactory and I resolved to develop the necessary logical

framework. Carrying out this resolve (Chapters 2 and 3) was the more difficult but the more valuable part of my research. A number of significant facts were uncovered that are not readily gleaned from the dynamic programming literature. For example, the principle of optimality need not be true for dynamic programming to be effective.

Several highly trained professors who use and teach dynamic programming have confided that they think dynamic programming is an art, and its use in solving a given physical or economic problem is more a matter of luck than science. Such an attitude must make the employment and exposition of dynamic programming a pretty dreary task. In order that the full power of the theory of multi-stage decision processes be available to engineers and statisticians, it is imperative that it first be developed as a mathematical subject, and that is the major objective here. The reader of this book will find that the principle of optimality and dynamic programming are natural parts of a novel, uncluttered, and highly challenging mathematical abstraction called "the adaptive control process."

SIDNEY J. YAKOWITZ

Tucson, Arizona
April, 1968

Acknowledgments: I am indeed grateful to my professors in the Electrical Engineering Department of Arizona State University for allowing me to work in an atmosphere of dignity and confidence while pursuing research interests which led me outside the traditional realms and viewpoints of engineering, and, to some extent, their capacity to evaluate. The Department allowed and even encouraged me to take half of my course work in the Mathematics Department. In particular, I am indebted to my thesis advisor, Dr. John D. Spragins, Jr., who introduced me to many tantalizing problems of engineering, and to Dr. Edward E. Grace (Mathematics Department), a topologist of the Moore tradition, who did as much as possible toward imparting the discipline of mind necessary to deal with these problems.

The Systems Engineering Department at the University of Arizona, where I now teach, has been generous in making available time, personnel, and facilities for assisting me in revising this manuscript. In particular I am grateful for the secretarial services of Mrs. Dee Ann Scott and Mrs. Dianne Felker and the programming assistance of Mr. William Freytag and Mr. Kinnear Williams.

As a graduate student I received support from a De Mund Foundation Graduate Fellowship and later an N.S.F. traineeship.

CONTENTS

CHAPTER I

Introduction

CHAPTER 2

Control Processes and Dynamic Programming

CHAPTER 3

Adaptive Control Processes

CHAPTER 4

Adaptive Control Processes in Two-Armed Bandit Theory

CHAPTER 5

Adaptive Control Processes in Pattern Recognition Theory

Appendices

Chapter I

INTRODUCTION

I. SOME REMARKS ON MATHEMATICAL MODELING

Chapter I stands entirely apart from the rest of the book. The concern of all chapters but this is the definition of abstract objects and, through rigorous deduction, the demonstration of consequences of these definitions. It is therefore appropriate that the later chapters are written from a mathematician's point of view. The purpose of this introductory chapter is to lay down some general principles for modeling so that problems arising in a physical or economic context may be viewed as adaptive control process problems. The intention here is to suggest rules for applying adaptive control process theory to human pursuits, and therefore we adopt an engineer's point of view.

The principal parts of a control process or an adaptive control process will be seen to be states, control elements, a law of motion, and a loss function, and this nomenclature is suggestive of the strict definitions that will come later.

When some worldly situation and set of desiderata are given, it is well to begin by reflecting on the desiderata. Decide what measurable quantities in the situation affect the cost of achieving the specified goal or desired output. Thus a sensible first step in abstracting an adaptive control process model from an actual situation is the specification of a *loss function* L. Next one should catalog the quantities on which L depends into two classes—those the designer is free to specify (collectively these will be called *controls*) and those the designer is not free to specify but which either depend on the controls or else are not within his sphere of influence. The variables of this second category will be termed *state coordinates*. For simplicity of analysis, one should remove from the state coordinates any variables which are redundant in the sense that they can be exactly calculated from other state coordinates. A specification of control and state variables is generally far from unique. In fact, in some models a given measurable quantity may appear as a control specification and in other models it may be a state coordinate.

The next task is for the designer to ascertain, as well as he is able, a law (to be called the *law of motion*) which specifies future states as a function (perhaps random) depending on the current state and future controls. This may generally be expected to be one of the most difficult steps in system analysis. It is accomplishments in system modeling which brought fame to Newton, Maxwell, and Einstein. It may be that the choice of state coordinates already made is not rich enough. That is, there may be measurements on which the loss does not depend, but which provide information necessary to calculate future states. The set of state coordinates should be enlarged to include such measurements.

Finally it is emphasized that the *state variable* (a vector of state coordinates) must be rich enough so that the future state may be calculated on the basis of present state and future input. This requirement is equivalent to the Markov property in probability theory. In the case that hysteresis effects are present in the system, some state coordinate must reflect hysteresis information.

If the above steps in modeling are successfully accomplished, the designer is in a position to make use of the facts and methods of adaptive control process theory. The designer is urged to perform experiments to make sure that predictions based on the model and adaptive control process analysis are fulfilled. Any conjectures about the law of motion should be treated as conjectures until verified by experiment. If for some reason experimentation is not practical and the system is not trivially simple, any deductions based on adaptive control process theory or any other mathematical study should be viewed with distrust and submitted with caution and humility.

2. THE NATURE OF CONTROL PROCESSES, ADAPTIVE CONTROL PROCESSES, TWO-ARMED BANDITS, AND PATTERN RECOGNITION MACHINES

The purpose of this section is to suggest the intuitive nature and the engineering importance of the models to be introduced in austere mathematical formality in subsequent chapters. All systems will have the property that for purposes of analysis, they exist only at discrete times (which may as well be taken to be the positive integers). The previous section remarked that the models are Markovian (or memoryless) in that future states are not dependent on the state and control history of the system. Often it is not difficult to modify non-Markovian models so that they have this property. Physical considerations have prompted adoption of the convention that all models are non-anticipatory in that motion does not depend on controls which have not yet been applied.

Our most simple system is the *control process* which, unlike the other models, is strictly deterministic. At each decision time j, the designer (or process-control computer) is to select a control element p from a set $P(x, j)$ of allowable controls which depend on the current state x as well as the decision time j. As a result of "applying" control p, the system will be in state $y = T(x, p, j)$ at the next decision time, $j + 1$. T, the law of motion, is part of the control process specification. Newtonian systems, chemical plants, and in fact any system described by differential or difference equations and wherein the controls are adjusted only at discrete times (an inevitable situation if a digital computer is used) are modeled easily as control processes. The object of control process analysis is to specify some sequence of control elements which cause the system to fulfill some objective at a minimum of expense.

It is unreasonable to model gaming, economic, communication, and feedback control systems as deterministic processes because, in interesting problems the future state is not a known function of the current state and control element. The *adaptive control process* differs from the control process model in assuming that the future state is a random variable related to the current state and control only through its probability law. The object of analysis is now to find a strategy which causes the system to fulfill some objective at a minimum of *expected* expense. It is not difficult to see that the adaptive control process is a suitable mathematical model for a great many popular multi-stage decision problems in game theory, economics, and communication theory wherein the statistical parameters are known. But the adaptive control process, by virtue of statistical decision theory, is also an appropriate model for situations demanding control in the absence of complete statistical knowledge. Under this circumstance, an adaptive control process solution specifies how to balance the cost of control and the expense of statistical estimation to achieve an overall minimum expected loss.

An adaptive control process with incomplete statistical knowledge which is the subject of an entire chapter is the *two-armed bandit*, a model that has received considerable scrutiny in the statistical literature. In two-armed bandit problems, we have two machines, I and II. At each play, I pays 1 or 0 with probability r or $1 - r$, respectively. II is the same as I, except it has a different Bernoulli parameter, s. It is presumed that the designer does not know the values r and s, and his object is to choose between machines sequentially so as to maximize his net take.

The two-armed bandit problem is a prototype of a sequential decision problem arising often in operations research. The usual approach to such problems is to have an experimental period (of fixed sample size) during which the unknown statistical parameter is estimated, using some classic technique. Then a decision is made as though the estimated value were known to be the true parameter value. An alternative approach suggests itself in two-armed bandit analysis. Through sequential design, as in the two-armed bandit technique, one can continue to improve performance after the formal experimental period, and often it would seem wiser to dispense with the sampling period altogether. In brief, with many problems arising in the study of economic processes, it seems unnatural and costly to separate the statistical analysis and control.

The situation in two-armed bandit analysis is essentially different from the discipline called "sequential analysis" studied by Wald [1] and others in that in the latter case there is no question of choosing between experiments (bandits). Only one experiment is available, and the decision is whether to continue testing or not.

The final model considered in this volume is the *pattern recognition machine*. The operation of the pattern recognition machine is roughly as follows: At each time in a discrete set, a pattern w (chosen randomly from a finite set) is presented to a transducer (a grid of photocells, a communication channel, etc.) which makes a noisy measurement z on it. On the basis of the measurement z, a computer attempts to "guess" the pattern w. It may be seen that this is the situation in Shannon's information theory for discrete noisy channels [2], although the questions of interest to us diverge from the object of Shannon's studies. By virtue of Shannon's sampling theorem [3], the problem of designing linear smoothing filters, studied by Wiener [4], can also be phrased in pattern recognition machine context if it is assumed that the transmitted signals are composed of strings of function segments chosen from a finite set.

Both Shannon and Wiener assumed in their studies that the channel statistics are known. In the past several years, communication theorists have begun studying the case (learning theory) in which this is not so. In our pattern recognition machine analysis, we show how a supervised learning period should be used to optimize performance. The optimization theory for pattern recognition machines is developed in a generality that makes it applicable to many current problems in communication and control.

The problem sections are intended to supplement this chapter in suggesting the engineering and statistical relevance of the mathematical theory in the remaining chapters.

REFERENCES

1. Wald, A. (1950). *Statistical Decision Functions*. Wiley, New York.
2. Shannon, C. (1948). A mathematical theory of communication. *Bell System Tech. J.* 27, 379–423, 623–656.
3. Oliver, B., J. Pierce, and C. Shannon (1948). The philosophy of PCM. *Proc. IRE 36,* 1324f.
4. Wiener, N. (1949). *Extrapolation, Interpolation and Smoothing of Stationary Time Series with Engineering Applications*. Wiley, New York.

Chapter 2

CONTROL PROCESSES AND DYNAMIC PROGRAMMING

I. DEFINITION OF A CONTROL PROCESS

In this chapter we are intent on modeling and analyzing a system having physically measurable, real-valued characteristics, some of which vary in time. Of these time-varying characteristics, there are only a certain finite number, n, which are relevant to the designer's interests. Thus an n-tuple of real numbers can display all desired measurements on a system at a fixed point in time. Such an n-tuple is called a *state*.

DEFINITION. *State-space* Ω is a subset of Euclidean n-space, R^n. A *state* x is an element of Ω.

It is supposed that the system we wish to model has the further structure that at each time t of a countable set A, if the system is in state x, a set of control elements $P(x, t)$ is available. At the discretion of the designer, some element $p \in P(x, t)$ is to be selected as the control element for state-time (x, t).

DEFINITION. The *set of decision times* A is the set of positive integers, or an initial segment thereof.

DEFINITION. The *control set function* P is a set-valued mapping defined on $\Omega \times A$, the set of state-time pairs. $P(x, t)$ is the *control set for* (x, t).

In control process theory (in contrast to adaptive control process theory), the system is assumed to behave deterministically. The designer knows what the state y at time $t + 1$ will be if he applies control element $p \in P(x, t)$ to the system at state-time (x, t). Mathematically speaking, the designer has a function T such that $y = T(x, p, t)$.

DEFINITION. A *law of motion* is a function T of the variable (x, p, t), where $p \in P(x, t)$. T takes on values in state space Ω.

6

DEFINITION. A *policy* $\bar{p}$ is a set of control elements $\{p(i)\}_{t \leqslant i \leqslant N}$ indexed by i, a variable on the domain indicated. N is the largest number in A, or ∞ in the absence of such a number.

In this book, indexed sets are considered functions on the domain of the index variable.

DEFINITION. Associated with a policy $\bar{p}$ and a state-time pair $(x(t), t)$ is an indexed set of states $\bar{x} = \{x(i)\}_{t \leqslant i \leqslant N+1}$ called a *trajectory*, such that $x(j + 1) = T(x(j), \bar{p}(j), j), t \leqslant j \leqslant N.$

The domain of definition of a policy (or its index variable) has one less element than its associated trajectory, since no decision is made at the terminal state.

The system we wish to model calls for one final abstract object. Intuitively, given some initial state-time, the designer perfers some policies to others. To model this situation, we suppose there is a real-valued function L which imposes an ordering such that $L(x, \bar{p}_a, t) < L(x, \bar{p}_b, t)$ means $\bar{p}_a$ is preferred to $\bar{p}_b$ at (x, t).

DEFINITION. A *loss function* L is a real-valued function defined on the variable $(x, \bar{p}_t, t)$. $\bar{p}_t$ is a policy defined for times not less than t.

If $\bar{p}$ is an arbitrary policy and we write $L(x, \bar{p}, t)$, we mean $\bar{p}$ restricted to times not less than t. In modeling in this manner, we imply that the designer considers his future cost to depend only on his present state and future controls.

DEFINITION. A *control process* is a quintuple (Ω, A, P, T, L), where Ω is a state space, A a set of decision times, P a control set function, T a law of motion, and L a loss function.

Many physical phenomena developing in time which can be influenced in a predictable manner may be modeled as control processes.

EXAMPLE 2.1. Consider a rocket whose thrusts can be adjusted only at millisecond intervals. Assume a set of Cartesian coordinate vectors $\bar{u}_1, \bar{u}_2, \bar{u}_3$ have been chosen. If x_j and v_j denote the position and speed, respectively, with respect to $\bar{u}_j, j = 1, 2, 3$, then the state vector is conveniently taken to be $x = (x_1, x_2, x_3, v_1, v_2, v_3)$ and the state space $\Omega = R^6$. If it is not specified when the flight must terminate, A may be taken to be the set of positive integers. $k \in A$ denotes the k-th decision time, which is also the k-th millisecond of flight. Assuming the thrust in any direction is bounded by M, then for all state-times (x, t), the control set $P(x, t)$ is $[-M, M] \times [-M, M] \times [-M, M]$ ($[-M, M]$ being the closed interval from $-M$ to M). The equation of motion comes from Newton's laws: If $x' = T(x, p, t)$, where $x = (x_1, x_2, \ldots, v_3)$ and $x' = (x_1', x_2', \ldots, v_3')$, then for $j = 1, 2, 3$,

$$v_j' = v_j + p_j \, 10^{-3},$$

$$x_j' = x_j + v_j \, 10^{-3} + \tfrac{1}{2} p_j \, 10^{-6}.$$

Imagine that the designer wishes to conserve fuel, and if $p(t)$ is the control at time t, then the fuel consumed during the t-th millisecond is $\|p(t)\|$, the Euclidean norm of $p(t)$. Then

$$L(x, \bar{p}, k) = \sum_{t \geqslant k} \|p(t)\|.$$

Since a state includes all the important information about a system, if the system is to accomplish any special objective, this objective may be specified in terms of conditions on state-space, and information about when these conditions are to be fulfilled. In general, these conditions will include a starting point—that is, an initial state-time pair. The goal in selecting a policy is to satisfy all the specified conditions and, at the same time, minimize loss.

DEFINITIONS.   The *conditions on state-space*, $\{Q(i)\} = Q$, is a set indexed by the same variable i which indexes trajectories; that is, i takes on values in $\{t, t + 1, \ldots, N + 1\}$. The elements of the range of Q are subsets of Ω. The conditions on state-space are *satisfied* by a trajectory $\bar{x}$ if, for each i, $\bar{x}(i) \in Q(i)$.

DEFINITION.   A *control process problem* is a control process and a set Q of conditions on state-space.

DEFINITION.   A *solution* to a control process problem is a policy $\bar{p}$ which generates a trajectory that satisfies the conditions on state-space and has the further property that, if (x, t) is the initial state-time,

$$L(x, \bar{p}, t) \leqslant L(x, \bar{p}', t)$$

for all policies $\bar{p}'$ whose trajectories satisfy the conditions on state-space.

EXAMPLE 2.1, CONTINUATION.   Referring to Example 2.1, we might specify that the rocket leave Cape Kennedy and land on the moon within ten days. This provides a control process problem. If (x_1, x_2, x_3), (y_1, y_2, y_3) are the coordinates of the Cape and the moon, respectively, and N the number of milliseconds in ten days, then the conditions on state-space are:

$$Q(1) = \{(x_1, x_2, x_3, 0, 0, 0)\}, \qquad Q(N) = \{(y_1, y_2, y_3, 0, 0, 0)\},$$

and

$$Q(t) = R^6, \qquad 1 < t < N.$$

A solution to this problem will be a policy whose trajectory satisfies the conditions $\{Q(i)\}$ with the least possible fuel consumption. The object of the next section is to derive an efficient method for finding such a solution.

Exercises in Control Process Modeling

(1) Find a control process model for the following situation: A caterer serves dinner for n_j people on the j-th night, $1 \leqslant j \leqslant N$. He can buy napkins at a cost of c_1 each, and have them cleaned by the next service at a cost of c_2/napkin by a fast cleaner. Also available is a slow cleaning service which does not have the napkins ready until the third night after they are used. The slow cleaner's rates are c_3/napkin. Finally, the caterer may rent napkins for any meal at a cost of c_4/napkin. He wishes to minimize his expenditures over the N-day period of operation.

(2) Find a control process model for the following situation: An irrigation district is served by two cylindrical reservoirs of areas A_1 and A_2 square meters, respectively. The reservoir system must supply g_i cubic meters of water on the i-th day, $1 \leqslant i \leqslant N$. The expense of an allocation scheme is reckoned to be

$$\sum_{i=1}^{N} a_i^2 + \sum_{i=1}^{N} b_i^3,$$

where a_i and b_i are the drop in water level of the first and second reservoir, respectively, on the i-th day.

(3) Model the irrigation problem above if the delivery requirement is only that at the end of the N-day period, G cubic meters of water have been delivered.

(4) Model the rocket problem discussed in Example 2.1 with the added restriction that the total fuel is V, and the object now is to reach the moon in minimum time rather than to conserve fuel. (*Hint:* The loss function and state variable of the example must be modified.)

(5) Find a control process model for the following situation: A manufacturing concern uses n(j) items of a certain type on the j-th day of operation, $1 \leqslant j \leqslant N$. These items cost q apiece; in addition there is a fixed service charge C for each order. Delivery is made before operation begins the day after the order is placed. Also there is a daily storage cost of s for each item in inventory at the end of a day. The manufacturer wishes to order the aforesaid items at a minimum total expenditure.

(6) Assume that a control process, a policy, and an initial state and time have been specified. Prove, using the principle of finite induction, that the associated trajectory may be computed.

(7) An investor with an initial capital C(1) and a daily utility* function u(w) has, at each day j, the option of spending any amount w(j) of C(j), his capital at that day, for his immediate enjoyment, which will have value $u(w(j))$ to him. The remaining capital, C(j) − w(j), is invested and will bring his capital to $Q(C(j) − w(j)) = C(j + 1)$ by the next day. The investor wishes to spend money at a rate which will maximize his enjoyment of an N-day period. Model this situation as a control process [3, p. 46].

(8) Given an m by n matrix A and vectors $C = (c_1, c_2, \ldots, c_n)$ and $B^T = (b_1, b_2, \ldots, b_m)$, where $b_i \geqslant 0$ for $1 \leqslant i \leqslant m$ and T indicates "transpose," the *standard maximum linear programming problem* is to find any or all non-negative $X^T = (x_1, x_2, \ldots, x_n)$ which maximize CX subject to the condition that

$$\text{i-th component (AX)} \leqslant \text{i-th component (B)}, \quad 1 \leqslant i \leqslant m.$$

Model the standard maximum linear programming problem as a control process problem. For a concise introduction to linear programming theory, see E. D. Nering, *Linear Algebra and Matrix Theory*, Wiley, New York (1963), pp. 197–208.

2. DYNAMIC PROGRAMMING

A certain mild restriction on the nature of the loss function causes the control process to possess a mathematical structure of great computational significance. Bellman [1, p. 54] describes the loss function restriction heuristically:

After any number of decisions, say k, we wish the effect of the remaining N-k stages of the decision process upon the total return to depend only upon the state of the system at the end of the kth decision and the subsequent decisions.

In essence, if two policies $\bar{p}_a$ and $\bar{p}_b$ are identical up to their k-th elements, and the elements after the k-th in $\bar{p}_a$ secure a lower loss than the remaining elements of $\bar{p}_b$ in getting from the k-th state to the end of the trajectory, then the restriction would ensure that L indicate $\bar{p}_a$ is the better policy.

The following definition expresses this restriction mathematically:

---

* The notion of utility is presented nicely in [9].

DEFINITIONS.   L is a *type* M *loss function* if it has the property that whenever

$$L\big(x(k), \bar{p}_a, k\big) < L\big(x(k), \bar{p}_b, k\big)$$

and

$$\bar{p}_a(i) = \bar{p}_b(i), \qquad j \leqslant i < k \tag{2.1}$$

then

$$L\big(x(j), \bar{p}_a, j\big) < L\big(x(j), \bar{p}_b, j\big). \tag{2.2}$$

A *Markovian control process* is a control process having a type M loss function.

REMARK.   If L is type M and the trajectories and policies are related as in the above definition, then

$$\big(L(x(j), \bar{p}_a, j) \leqslant L(x(j), \bar{p}_b, j)\big) \Rightarrow \big(L(x(k), \bar{p}_a, k) \leqslant L(x(k), \bar{p}_b, k)\big). \tag{2.3}$$

EXAMPLE 2.2.

$$L\big(x(t), \bar{p}, t\big) = \left(\sum_{i \geqslant t} |a_i x(i)|^{\alpha_i} + |b_i \bar{p}(i)|^{\beta_i} \right)^{\mathrm{v}}, \tag{2.4}$$

where a_i, b_i, α_i, and β_i are real, and v is positive. The existence of the law of motion T assures that $x(i)$, $i > t$ can be calculated from $\bar{p}$ and $x(t)$. In fact, $x(t + 1) = T\big(x(t), \bar{p}(t), t\big)$, $x(t + 2) = T\big(x(t + 1), \bar{p}(t + 1), t + 1\big)$, etc. To show L above is type M, in the notation of the definition, let

$$c = \sum_{i=j}^{k-1} |a_i x(i)|^{\alpha_i} + |b_i \bar{p}(k)|^{\beta_i},$$

$$c_1 = L\big(x(k), \bar{p}_a, k\big) < L\big(x(k), \bar{p}_b, k\big) = c_2.$$

Then

$$L\big(x(j), \bar{p}_a, j\big) = (c_1^{1/\mathrm{v}} + c)^{\mathrm{v}} < (c_2^{1/\mathrm{v}} + c)^{\mathrm{v}} = L\big(x(j), \bar{p}_b, j\big),$$

since u^{v} is a monotonically increasing function of u. Loss functions of the above form particularly popular in the engineering literature are $\sum |p(i)|^2$, $\sum |p(i)|$, and $\big(\sum |p(i)|^2\big)^{1/2}$.

EXAMPLE 2.3.   For $x, y \in \Omega$, $p \in P(x, \zeta)$, $\zeta \in A$, let $\rho(x, y, p, \zeta)$ be any real-valued function of the indicated variables. Then

$$L(x, \bar{p}, t) = \sum_{i \geqslant t} \rho\big(x(i), x(i + 1), \bar{p}(i), i\big) \tag{2.5}$$

is a type M loss function. To demonstrate this assertion, let

$$c = \sum_{i=j}^{k-1} \rho\big(x(i), x(i + 1), \bar{p}(i), i\big).$$

If $L\big(x(k), \bar{p}_a, k\big) < L\big(x(k), \bar{p}_b, k\big)$, then

$$L\big(x(j), \bar{p}_a, j\big) = c + L\big(x(k), \bar{p}_a, k\big) < c + L\big(x(k), \bar{p}_b, k\big)$$
$$= L\big(x(j), \bar{p}_b, j\big).$$

Loss functions representable in the form (2.5) will be called *separable*. A variation on the separable loss function is:

$$L(x, \bar{p}, t) = \varphi\left(\sum_{i \geq t} \rho\big(x(i), x(i+1), \bar{p}(i), i\big)\right), \qquad (2.6)$$

where φ is a strictly increasing real monotonic function, and ρ is as above. This function is also type M. For if

$$c = \sum_{i=j}^{k-1} \rho\big(x(i), x(i+1), p(i), i\big)$$

and $\bar{p}_a$, $\bar{p}_b$, $x(j)$, and $x(k)$ are related as in the definition and

$$L\big(x(k), \bar{p}_a, k\big) < L\big(x(k), \bar{p}_b, k\big),$$

then

$$c_1 \equiv \varphi^{-1}\big(L(x(k), \bar{p}_a, k)\big) < \varphi^{-1}\big(L(x(k), \bar{p}_b, k)\big) = c_2,$$

which implies

$$L\big(x(j), \bar{p}_a, j\big) = \varphi(c + c_1) < \varphi(c + c_2) = L\big(x(j), \bar{p}_b, j\big).$$

EXAMPLE 2.4. The loss function below is a curiosity because it is a type M loss function which is not a monotonic function of a separable loss function. Let v be any positive integer:

$$L\big(x(t), \bar{p}, t\big) = \sum_{i=t}^{v-1} |x(i)| + \left(\sum_{i \geq v} |x(i)|^2\right)^{1/2}.$$

To show this is type M, first consider the case that k (in the definition of type M functions) is less than v and

$$c = \sum_{i=j}^{k-1} |x(i)|.$$

Then $L\big(x(k), \bar{p}_a, k\big) < L\big(x(k), \bar{p}_b, k\big)$ implies

$$L\big(x(j), \bar{p}_a, j\big) = c + L\big(x(k), \bar{p}_a, k\big) < c + L\big(x(k), \bar{p}_b, k\big) = L\big(x(j), \bar{p}_b, j\big).$$

Alternatively, if $k \geq v$, and

$$c = \sum_{i=j}^{v-1} |x(i)|, \qquad \text{and} \qquad c_1 = \sum_{i=v}^{k-1} |x(i)|^2,$$

then

$$c_a \equiv L\big(x(k), \bar{p}_a, k\big) < L\big(x(k), \bar{p}_b, k\big) \equiv c_b$$

implies

$$L\big(x(j), \bar{p}_a, j\big) = c + \big(c_1 + (c_a)^2\big)^{1/2} < c + \big(c_1 + (c_b)^2\big)^{1/2} = L\big(x(j), \bar{p}_b, j\big).$$

DEFINITIONS. Assume a control process problem has been specified, and $k \in A$ and $x \in Q(k)$. Then $\{Q(i)\}_{i \geq k}$ are the conditions for the *modified problem at* (x, k). For this modified problem, x is the initial state and k the initial decision time. A policy $\bar{p}$ is a *solution to the modified problem at* (x, k) if the trajectory $\bar{x}$ associated with $\bar{p}$ satisfies the conditions of the modified problem (i.e., $\bar{x}(i) \in Q(i)$, $i \geq k$), and if $\bar{p}'$ is any other policy whose trajectory also satisfies the conditions of the modified problem,

$$L(x, \bar{p}, k) \leq L(x, \bar{p}', k).$$

Many problems arising in engineering, economics, operations research, and physics are abstracted naturally as Markovian control process problems. A solution to a Markovian control process problem has an important property described in the following theorem:

THEOREM 2.1 (Principle of Optimality): *Assume $\bar{p}$ is a solution to a Markovian control process problem with initial state-time $\big(x(j), j\big)$ and $\bar{x}$ is the trajectory generated by $\bar{p}$. Then $\bar{p}$ is also a solution to the modified problem at $\big(\bar{x}(k), k\big)$ for every $k \in A$ such that $k > j$.*

The theorem above is intended to be a precise statement of the assertion Bellman [1, p. 57] denotes "the principle of optimality," which is:

An optimal policy has the property that whatever the initial state and initial decision are, the remaining decision must constitute an optimal policy with regard to the state resulting from the first decision.

PROOF. Let $\bar{p}'$ be any policy satisfying the conditions of the modified problem at $\big(\bar{x}(k), k\big)$. Construct the policy $\bar{p}''$ so that

$$\bar{p}''(i) = \begin{cases} \bar{p}'(i), & i \geq k, \\ \bar{p}(i), & i < k. \end{cases}$$

Since $\bar{p}$, $\bar{p}'$ both generate trajectories satisfying the conditions on state-space, the trajectory associated with $\bar{p}''$ also satisfies the conditions on state-space. The hypothesis that $\bar{p}$ is a solution to the original problem gives

$$L\big(x(j), \bar{p}, j\big) \leq L\big(x(j), \bar{p}'', j\big). \tag{2.7}$$

Observe that $\bar{p}$ and $\bar{p}''$ are related as $\bar{p}_a$, $\bar{p}_b$ in the definition of "type M" function. By (2.3), (2.7) implies

$$L(\bar{x}(k), \bar{p}, k) \leqslant L(\bar{x}(k), \bar{p}'', k). \tag{2.8}$$

As $\bar{p}'$ restricted to times not less than k is identical to $\bar{p}''$,

$$L(\bar{x}(k), \bar{p}', k) = L(\bar{x}(k), \bar{p}'', k).$$

In summary,

$$L(\bar{x}(k), \bar{p}, k) \leqslant L(\bar{x}(k), \bar{p}', k). \tag{2.9}$$

As $\bar{p}'$ was selected arbitrarily from policies satisfying the conditions of the modified problem, (2.9) implies the theorem.

The preceding theorem points out the interesting "principle of optimality" property that a solution has, namely, that it is a solution to all modified problems occurring along the process trajectory. Incidentally, this proves that those particular modified problems have solutions.

The principle of optimality theorem allows us to conclude that in searching for a solution to a problem with initial state-time (x, j) it is sufficient to consider only those policies which are solutions to modified problems at time $j + 1$. The fundamental theorem to follow leads to a procedure for finding a solution from the class of policies that are solutions to the modified problems.

THEOREM 2.2. (Dynamic Programming Theorem): *Suppose a Markovian control process problem with initial state-time* $(x(j), j)$ *has a solution, and the function* $\bar{v}(y)$ *is a mapping into a solution to the modified problem at* $(y, j + 1)$ *at those states* $y \in Q(j + 1)$ *at which solutions exist. Define*

$$0 = \{p: p \in P(x, j) \quad and \quad T(x, p, j) \in Q(j + 1)\}$$

and for each $p \in 0$, *let* $\hat{p}$ *be the policy defined*

$$(i)\hat{p} = \begin{cases} p, & if\ i = j, \\ \bar{v}(T(x, p, j))(i), & if\ i > j. \end{cases}$$

If $\hat{p}^*$ *is a policy such that* $p^* \in 0$ *and*

$$L(x, \hat{p}^*, j) = \min_{p \in 0} L(x, \hat{p}, j),$$

then $\hat{p}^*$ *is a solution to the control process problem.*

PROOF. Let $\bar{p}$ be a solution to the problem, and observe that, by the principle of optimality theorem, $\bar{p}$ is a solution to the modified problem at

$(y, j + 1)$, where $y = T(x, \bar{p}(j), j)$. From the definition of $\bar{v}$ and the principle of optimality theorem, we have that $\bar{v}$ is defined at y and thus

$$L(y, \bar{p}, j + 1) = L(y, \bar{v}(y), j + 1). \qquad (2.10)$$

As $\bar{p}$ is a solution and therefore generates a trajectory satisfying conditions on state-space, we have that $\bar{p}(j) \in 0$. Thus (2.10) implies, by the fact that L is type M,

$$L(x, \bar{p}, j) = L\big(x, \widehat{\bar{p}(j)}, j\big). \qquad (2.11)$$

The proof is now concluded by observing that, from the definition of $\hat{p}^*$,

$$L(x, \hat{p}^*, j) = \min_{p \in 0} L(x, \hat{p}, j) \leqslant L\big(x, \widehat{\bar{p}(j)}, j\big) = L(x, \bar{p}, j). \qquad (2.12)$$

The dynamic programming theorem is the theoretical justification for the analytic methods which the expression "dynamic programming" is usually intended to denote. The extreme computational and analytic significance of dynamic programming will be suggested as soon as we reveal the dynamic programming method for Markovian control processes bounded in time.

ALGORITHM. *Dynamic Programming for Markovian Control Processes with $\bar{\bar{A}} = N < \infty$.*†

For each state $x \in Q(t)$ let $0(x, t) = \{p : p \in P(x, t)$ and $T(x, p, t) \in Q(t + 1)\}$. Let $\bar{v}(x, N) = p^*(x, N) \in 0(x, N)$‡ be a control such that

$$L\big(x, p^*(x, N), N\big) = \min_{p \in 0(x, N)} L(x, p, N). \qquad (2.13)$$

If the indicated minimum does not exist, then $\bar{v}$ will not be defined at (x, N). It is a consequence of Theorem 2.1 that, if the control process problem has a solution, then there must be some problem at time N for which the minimum exists. By construction, $\bar{v}$ is a solution at those problems where it is defined. We develop the function $\bar{v}(x, t)$ inductively on decreasing numbers t, where $\bar{v}$ maps the pair (x, t) into a solution to the modified problem at (x, t). Assume $\bar{v}$ has been constructed for $t = m + 1$. Then, for $x \in Q(m)$, $\bar{v}(x, m)$ is defined as $\bar{v}$ in Theorem 2.2, that is

$$\bar{v}(x, m)(i) = \begin{cases} p^*, & \text{if } i = m, \\ \bar{v}\big(T(x, p^*, m), m + 1\big)(i), & \text{if } i > m, \end{cases} \qquad (2.14)$$

† Here and elsewhere, "$\bar{\bar{s}}$" will denote "number of elements in the set s."

‡ The fastidious will notice that the image of $\bar{v}(x, N)$ should be a one-element policy, not a control element.

where p* is any element of $0(x, m)$ such that

$$L(x, \hat{p}^*, m) = \min_{p \in 0(x,m)} L(x, \hat{p}, m)$$

($\hat{p}$ is as defined in Theorem 2.2). By Theorem 2.2, $\bar{v}(x, m)$ is a solution to the modified problem at (x, m). In particular, if (x, j) is the initial state and time of a control process problem, then $\bar{v}(x, j)$ is a solution to that problem.

If the control process sets are finite, then dynamic programming (hereafter abbreviated DP) is a powerful computational scheme for obtaining solutions. Obviously, if a solution exists to the finite problem, it can be found in finitely many comparisons by testing all policies. Hence a method is not particularly interesting unless it produces a solution efficiently in some sense. It is in this context that the computational features of DP are notable because very often one can calculate an upper bound to the number of comparisons that must be made which is lower than the number associated with a total search. From the description of the DP algorithm, one can see that, for each pair (x, j), there are at most $\overline{\overline{0(x, j)}}$ comparisons required to determine $v(x, j)$. Often this fact leads to great economy over a search through all possible policies. For instance, if $\overline{\overline{0(x, j)}} = \overline{\overline{0(j)}}$, independent of x, then the total number of comparisons necessary to find a solution by DP is bounded by

$$\overline{\overline{\Omega}} \sum_{j=1}^{N} \overline{\overline{0(j)}}.$$

If all possible policies are tested,

$$\prod_{j=1}^{N} \overline{\overline{0(j)}}$$

comparisons are required.

For many researchers, the analytic aspects are the most important facet of DP. Without restricting attention to finite processes, it has been used to demonstrate the existence, uniqueness, certain structural features, and even the construction of process solutions. In Chapter 4 of this book, for instance, it is divulged by means of DP that, for a particular type of two-armed bandit problem, regardless of process parameters, once a particular control element p′ appears in a solution, all remaining elements are p′. Blackwell [2] and others have answered intriguing mathematical questions by placing them in a control process framework and using DP ideas.

The first example to follow illustrates the use of DP as a computational tool, and the second demonstrates its analytic merits.

EXAMPLE 2.5. (The Computational Aspects of DP): Mr. G sells hot watches. His fence will pay on any day \$log (n) for n watches, log (0) reflecting the dealer's wrath at no deliveries. How should G get rid of his stock of 10 watches in three days in order to maximize his income?

SOLUTION. The control p(t) at each day t, t = 1, 2, 3, is how many watches to dispose of. This cannot exceed $10 - x(t)$ where $x(t)$ is the number of watches sold thus far. Thus G wishes to choose $\bar{p} = \{p(i)\}_{i=1}^{3}$ so as to maximize

$$\sum_{i=1}^{3} \log\left(p(i)\right) = \log\left(\prod_{i=1}^{3} p(i)\right),$$

under the constraints that the p(i)'s be non-negative integers and $\sum_{i=1}^{3} p(i) \leqslant 10$. Because log (v) is a monotonically increasing function of v, the problem is equivalent to finding a strategy $\bar{p}$ which maximizes $\prod_{i=1}^{3} p(i)$.

Thus state-space is $\Omega = \{0, 1, \ldots, 10\}$, where x indicates the number of watches already disposed of. For each state-time pair (x, i), the control set

$$P(x, i) = \{1, 2, \ldots, 10 - x - (3 - i)\},\dagger$$

and the law of motion is $y = T(x, p, i) = x + p$. The loss function is

$$L(x, \bar{p}, j) = -\prod_{i \geqslant j} p(i).$$

It is clear that this loss function is type M for positive control elements. The initial state is 0.

For decision time 3, each policy $\bar{p}$ has but one control element, p. $L(x, \bar{p}, 3) = -p$. Hence the smallest loss is obtained by selecting as large a control element as possible. That is,

$$\overline{v(x, 3)} = \{\max P(x, 3)\} = \{10 - x\}.$$

Table 2.1 lists the control element in the solutions to all modified problems at (x, 3), $x \in \Omega$.

For modified problems at t = 2, policies are representable as 2-tuples $\bar{p} = \{p(2), p(3)\}$, where p(3) is the control of

$$\bar{v}(T(x, p(2), 2), 3) = \bar{v}(x + p(2), 3).$$

For each x, we enumerate $L(x, \hat{p}, 2) = -p(2)p(3)$ for all policies $\hat{p}$ as defined in Theorem 2.2. A policy $\hat{p}^*$ yielding the minimum loss is selected.

† We exclude from consideration the possibility that on any day 0 watches are delivered, since this policy would obviously be dominated (by (1, 1, 1), for instance).

**Table 2.1. Solutions and Associated Losses to Modified
Problems at Time 3**

State x	Solution $v(x, 3)$	Loss $L(x, \overline{v(x, 3)}, 3)$
9	1	−1
8	2	−2
7	3	−3
6	4	−4
5	5	−5
4	6	−6
3	7	−7
2	8	−8
1	9	−9
0	10	−10

By authority of Theorem 2.2, the Dynamic Programming Theorem, $\hat{p}^*$ will be a solution to the modified problem at $(x, 2)$.

For instance, if $x = 5$, then $P(x, 2) = P(5, 2) = \{1, 2, 3, 4\}$. If p is chosen to be 1, then the new state will be $x + p = 5 + 1 = 6$. In accordance with the definition of $\hat{p}$ (see Theorem 2.2),

$$\hat{1} = \left(1, \overline{v}(6, 3)\right) = (1, 4).$$

The loss for this policy is $-\prod_{i \geqslant 2} p(i) = -1 \cdot 4 = -4$. Continuing in this fashion for all elements in $P(5, 2)$, Table 2.2 is constructed.

**Table 2.2. Enumeration of Policies "$\hat{p}$" for the
Modified Problem at $(5, 2)$**

Control p	New state $T(5, p, 2)$	Policy $\hat{p}$	Loss $L(5, \hat{p}, 2)$
1	6	(1, 4)	−4
2	7	(2, 3)	−6
3	8	(3, 2)	−6
4	9	(4, 1)	−4

Examination of Table 2.2 reveals that $p(2) = 3$ may be taken for p^*, since no policy gives a lower loss for the modified problem at $(5, 2)$ than $(3, 2)$. By repeating this procedure for all states, the solutions to all modified problems at time 2 are found. They are given in Table 2.3.

**Table 2.3. Solutions and Associated Losses to
Modified Problems at Time 2**

State x	Control p*	New state T(x, p*, 2)	Loss L(x, p̂*, 2)	Solution v̄(x, 2)
8	1	9	-1	(1, 1)
7	2	9	-2	(2, 1)
6	2	8	-4	(2, 2)
5	3	8	-6	(3, 2)
4	3	7	-9	(3, 3)
3	4	7	-12	(4, 3)
2	4	6	-16	(4, 4)
1	5	6	-20	(5, 4)

Finally, solutions are found to the total problem by finding the best policy p̂ at state-time (0, 1). The policies p̂ are found as in Theorem 2.2, that is, p̂ $= \{p(i)\}_{i=1}^{3}$, where

$$\hat{p}(i) = \begin{cases} p, & \text{if } i = 1, \\ \overline{v(p, 2)}(i), & i = 2, 3. \end{cases}$$

$L(0, \bar{p}, 1) = -\prod_{j=1}^{3} p(j)$. All policies p̂ and associated losses are enumerated in Table 2.4.

Thus both (3, 4, 3) and (4, 3, 3) are solutions to Mr. G's watch allocation problem, and his stock, if one of these solutions is used, is worth \$log (36).

The purpose of the examples here is to give substance to control process theory, and so the exposition goes at an easy gait. With practice, the reader

**Table 2.4. Enumeration of Policies "p̂" and
Associated Losses for the Total Control
Process Problem**

Control p	New state T(0, p, 1)	Policy p̂	Loss $L(0, \hat{p}, 1) = \prod_{j=1}^{3} p(i)$
1	1	(1, 5, 4)	-20
2	2	(2, 4, 4)	-32
3	3	(3, 4 3)	-36
4	4	(4, 3, 3)	-36
5	5	(5, 3, 2)	-30
6	6	(6, 2, 2)	-24
7	7	(7, 2, 1)	-14
8	8	(8, 1, 1)	-8

will be able to do problems much more compactly. Nevertheless, it is doubt-ful that the computational aspects of DP would have gained such attention in the scientific community if digital computers were not available. The following example illustrates the power of DP as an analytic tool, and this power is independent of the existence of computers.

EXAMPLE 2.6.   (The Analytic Aspects of DP) (after [3], p. 41): For some positive number v, define the function

$$f_N(a) = \min_R \sum_{i=1}^N p(i)^v, \qquad a > 0, \tag{2.15}$$

where R is the region defined by

(a) $\sum_{i=1}^N p(i) \geqslant a$,

(b) $p(i) \geqslant 0$.

Show that $f_N(a) = a^v c_N$, where c_N depends only on N and v.

ANALYSIS.   Strangely enough, it is worthwhile to model this mathematical problem as a control process problem. We let the state be $x = \sum_{i \in A'} p(i)$, where A' is an initial segment of the domain of p̄. The control space consistent with the problem description of the region R is

$$P(x, j) = \begin{cases} [0, \infty), & \text{if } j < N, \\ [w, \infty), & \text{if } j = N, \end{cases} \tag{2.16}$$

where $w = \max\{0, a - x\}$. The law of motion is $T(x, p, i) = x + p$, and the initial state is 0. $L(x, \bar p, j) = \sum_{i \geqslant j} (p(i))^v$, which is separable and thus type M. If we construct the function v̄ by dynamic programming, then

$$f_N(a) = L(0, \overline{v(0, 1)}, 1). \tag{2.17}$$

We demonstrate the assertion of the example by induction on N. For $N = 1$, $p(0, 1) = [w, \infty) = [a, \infty)$. "$p(1) = a$" gives the smallest loss, and thus $f_1(a) = (p(1))^v = a^v$ and $c_1 = 1$, independent of a. Assume the assertion is true for $1 \leqslant N < m$. Then

$$f_m(a) = \min_{p \in [0, a]} [p^v + f_{m-1}(a - p)] = \min_{p \in [0, a]} [p^v + c_{m-1}(a - p)^v];† \tag{2.18}$$

† (2.18) makes use of the fact that, because $p' > a$ implies $(p')^v > (a)^v \geqslant f_m(a)$, the minimizing value p^* of $f_m(a)$ cannot exceed a.

the last equality uses the inductive hypothesis. We find the extremum of $H(p) = p^v + (a - p)^v c_{m-1}$, $p > 0$:

$$\frac{dH(p)}{dp} = vp^{v-1} - c_{m-1}v(a - p)^{v-1} = 0. \tag{2.19}$$

Since $d^2/dp^2(H(p)) \geqslant 0$, the unique value p^* satisfying (2.19) must be a global minimum. From (2.19), we have

$$(p^*)^{v-1} = c_{m-1}(a - p^*)^{v-1}$$

or

$$p^* = (c_{m-1})^q (a - p^*),$$

where $q = 1/(v - 1)$. Hence

$$p^* = [(c_{m-1})^q/1 + (c_{m-1})^q]a \equiv \eta a. \tag{2.20}$$

Substituting this minimizing value of p into (2.18), we have

$$f_m(a) = [\eta^v + c_{m-1}(1 - \eta)^v]a^v. \tag{2.21}$$

Thus

$$c_m = \left(\frac{(c_{m-1})^q}{1 + (c_{m-1})^q}\right)^v + c_{m-1}\left(1 - \frac{(c_{m-1})^q}{1 + (c_{m-1})^q}\right)^v$$

independently of a, and $f_N(a) = c_N a^v$ for all positive numbers N by authority of the first principle of induction.

In the preceding theorems it was postulated that solutions exist to the Markovian control process (CP) problems. Thus one would like weak conditions ensuring the existence of a solution. If all sets are finite and there is any policy generating a trajectory satisfying the problem conditions, then a solution exists. In the more general case, if for all decision times i, $O(x, i) = O(i)$, independent of x, where $O(i)$ is a compact set of control elements, then a solution exists provided the loss function is continuous and some policy satisfies the problem conditions. This statement is a consequence of the fact that by Tychonoff's theorem [5, p. 143], the space of policies being the Cartesian product of compact control sets, is compact. As L is a continuous function of the compact space of policies, L attains its maximum at some policy [6, p. 73].

At present, there seem to be no general existence statements. The difficulty stems from the fact that we have made the control sets dependent on state. Many systems such as inventory processes seem to call for such a model. Bellman [3] has established the existence of solutions to several important types of control process problems.

Exercises on Dynamic Programming for Control Processes

(1) Solve the caterer's problem (Problem 1, Section 1) with the customer sequence $(3, 5, 5, 2)$ and the parameters $c_1 = 10$, $c_2 = 3$, $c_3 = 1$, and $c_4 = 5$. Assume the caterer is indifferent to whether or not a napkin is clean at the end of the four day service. (For a discussion of this sort of problem, see [4].)

(2) Consider the problem of finding a policy $\bar{p}$ which attains the value

$$f_N(a) = \max_{\bar{p} \in D} \prod_{i=1}^{N} p(i)$$

where D is the region defined by the relations $p(i) \geqslant 0$ and $\sum_{i=1}^{N} p(i) \leqslant a$. Model this problem as a control process problem and solve it. Find a general expression for the value of $f_N(a)$ [3, p. 40].

(3) A traveling salesman is told to visit N cities $\{c_1, c_2, \ldots, c_N\}$ and provided with a map which displays the distance between them. Let d_{ij} represent the distance between c_i and c_j. The salesman wishes to find a tour that will include all cities and minimize the distance traveled. Model this problem as a control process problem and find an upper bound to the number of comparisons which must be made. What is the total number of tours beginning at c_1 wherein no city is visited twice?

(4) $g_j(x) = 1 - x^j$, $j = 1, 2, 3$. By modeling as a control process problem, find

$$\max_{\bar{p} \in D} \sum_{i=1}^{3} g_j\big(p(i)\big)$$

where D is the region in R^3 defined by $p(i) \geqslant 0$ and $\sum_{i=1}^{3} p(i) = 1$.

(5) Indicate formally (e.g., by writing a computer flow chart or program) how to solve one of the problems 2, 3, 4, 5, or 7 of the problem set in Section 1.

(6) For one of the problems 2, 3, 4, 5, or 7 of Section 1, make some non-trivial numerical specification for the process parameters and solve, using a digital computer to perform the dynamic programming algorithm.

(7) There are N different types of items, with the i-th item having weight w_i and value v_i. It is desired to load a ship having total capacity of w pounds with a cargo of the greatest possible value. Show that this problem leads to the problem of determining the maximum over $\bar{n} = \{n(i)\}_{i=1}^{N}$ of the linear form

$$L(\bar{n}) = \sum_{i=1}^{N} n(i)v_i,$$

subject to the constraint that the $n(i)$'s be non-negative integers and $\sum_{i=1}^{N} n(i)w_i \leqslant w$. Model this situation as a control process problem and indicate formally an efficient algorithm for finding a solution [3, p. 45].

3. HISTORY AND SUMMARY OF CONTROL
PROCESS THEORY

The first comprehensive publication on control process theory and dynamic programming is *Dynamic Programming* by Bellman [3], which is still the standard reference for these subjects. Bellman is the mathematician who coined the term "dynamic programming." The nature of this book is exploratory rather than axiomatic. It is seen that interesting problems arising in economics and game theory are profitably studied by observing their control process structure. Thus the emphasis of this volume is on demonstrating the importance of control process models to economic situations rather than the logical structure of control processes themselves. Each economic problem is tackled on its own ground. But, by example, Bellman brings out the pattern of analysis. This is a pioneering work in two respects: It observes the common mathematical structure of apparently diverse problems, and it demonstrates the economy and elegance of the dynamic programming algorithm in analyzing these problems. *Dynamic Programming* contains an excellent bibliography of the research papers preceding its publication.

Karlin's paper [7] is widely recognized to be a fundamental contribution to the analysis of the logical structure of control processes, the subject of this chapter. Differences between his development and ours are that he considers only separable loss functions and does not allow control sets to be indexed by states. He postulates the state and control sets to be arbitrary topologic spaces.

Mitten [8] recognized that the dynamic programming algorithm works with more general loss functions than separable functions. He proved that that the dynamic programming algorithm produces a solution if L has the "monotonicity property," which, in our notation from the definition of type M functions, is

$$\big(L(x(k), \bar{p}_a, k) \leqslant L(x((k), \bar{p}_b, k)\big) \Rightarrow \big(L(x(j), \bar{p}_a, j) \leqslant L(x(j), \bar{p}_b, j)\big). \quad (2.22)$$

This differs from "type M" in that the strong inequalities have been reduced to weak inequalities. Thus the set of monotonic functions subsumes type M loss functions. In the monotonicity framework, however, the principle of optimality need not be true.

The significance of the monotonicity property is clarified in Denardo's dissertation [10] (written under Mitten's guidance), which presents a far-reaching theory for sequential decision processes. A number of problems in the dynamic programming literature are shown to be within the scope of this theory.

Nemhauser's book [15] gives a logical development of the material of this chapter at a more leisurely pace. It includes Mitten's results on monotonicity.

This chapter ignores the application of dynamic programming techniques. Bellman and Dreyfus [11], Bellman and Kalaba [12], Dreyfus [13], and Hadley [14] will serve to introduce the interested reader to this vast subject. In particular, many results of variational calculus can be derived by considering the behavior of control processes as the time between decisions converges to zero [13].

Electrical engineers often use the designation "discrete optimal control process" for what we call "control process." By "control process" they usually mean a statistical system (random disturbances), although probability theory seldom appears in their analyses. The analysis generally seeks to attain a qualitative standard of performance, and a loss function is seldom explicitly stated. Fundamentally, electrical engineering control processes are more closely related to adaptive control processes, the subject of the rest of this book.

Problems on the Monotonicity Property

1. Let $\rho(x, y, p, i)$ be as in the definition of a separable loss function. Show that

$$L(x, \bar{p}, i) = \min_{n \geq i} \{\rho(x, y, p(n), n)\}$$

is a loss function with the monotonicity property.

2. Does

$$L(x, \bar{p}, t) = \left(\sum_{i=t}^{m} |x(i)|\right)^{1/2} + \left(\sum_{i>m} |x(i)|\right)^{1/3}$$

have the monotonicity property?

REFERENCES

1. Bellman, R. (1961). *Adaptive Control Processes. A Guided Tour.* Princeton University Press, Princeton, N.J.
2. Blackwell, D. (1964). Probability bounds via dynamic programming. *Proc. Symp. Appl. Math. 16*, 277–289.
3. Bellman, R. (1957). *Dynamic Programming.* Princeton University Press, Princeton, N.J.
4. Elmagrhabi, S. (1964). Dynamic programming approaches to the caterer's problem. *J. Math. Anal. Appl. 8*, 202f.

5. Kelley, J. L. (1955). *General Topology*. Van Nostrand, Princeton, N.J.
6. Apostol, T. M. (1957). *Mathematical Analysis*. Addison-Wesley, Reading, Mass.
7. Karlin, S. (1955). The structure of dynamic programming models. *Naval Res. Logist. Quart. 2*, 285–294.
8. Mitten, L. (1964). Composition principles for synthesis of optimal multi-stage processes. *Operations Res. 12*, 610–619.
9. Luce, R. D., and H. Raiffa (1957). *Games and Decisions*. Wiley, New York.
10. Denardo, E. (1965). *Sequential Decision Processes*. Dissertation, Northwestern University. University Microfilms No. 65-12,069.
11. Bellman, R., and S. Dreyfus (1962). *Applied Dynamic Programming*. Princeton University Press, Princeton, N.J.
12. Bellman, R., and R. Kalaba (1966). *Dynamic Programming and Modern Control Theory*. Academic Press, New York.
13. Dreyfus, S. E. (1965). *Dynamic Programming and the Calculus of Variations*. Academic Press, New York.
14. Hadley, G. (1964). *Nonlinear and Dynamic Programming*. Addison-Wesley, Reading, Mass.
15. Nemhauser, G. L. (1966). *Introduction to Dynamic Programming*. Wiley, New York.

Chapter 3

ADAPTIVE CONTROL PROCESSES

I. DEFINITION OF AN ADAPTIVE CONTROL PROCESS

If the behavior of a control process is random in some situations, then interesting statistical possibilities arise. The processes studied in the remainder of this book are that small portion of probabilistic control process generalizations which result when the law of motion in state-space becomes a random experiment. Specifically there is at least one state-time pair (x, t) and one control element p such that, if the system is in state x at time t and control p is applied, the resulting state will not be known in advance. Instead, prior to the movement actually taking place, the future state is a random variable (r.v.) Y described by the n-dimensional cumulative distribution function (c.d.f.) $F(y \mid x, p, t)$ indexed by the state-time pair (x, t) and control p. We shall use capital letters to denote r.v.'s and lower case letters to denote observations of those r.v.'s. The class of systems modeled here further requires that, after the movement has taken place, knowledge of y, the resulting state, be made available to the decision mechanism for consideration in selecting the control element for time $t + 1$. This requirement is equivalent to "feedback" in electrical engineering control theory. Throughout this book, *all integrands will be assumed integrable with respect to the indicated measure.* When no range of integration is indicated, it is assumed to be the entire domain of the variable of integration.

The stochastic regime we are considering here calls for major modifications of the control process model. The notions of state-space, control set function, and decision times are applicable unchanged. For the deterministic law of motion is substituted a family $\mathscr{F}$ of c.d.f's indexed by the variable (x, p, t), where $x \in \Omega$, $p \in P(x, t)$, and $t \in A$. The family $\mathscr{F}$ will be called the *statistical law of motion.* In the determininstic case, if the policy and initial state-time are specified, no further information is gained by specifying also the trajectory, since this is a given function of policy and initial state-time. In the statistical process, the trajectory cannot be reconstructed from the policy and initial state-time, and, to achieve generality, the loss function is made to depend on the trajectory as well as the policy.

26

Finally, the concept of policy has only limited usefulness in the statistical regime; it is no longer a satisfactory specification for guiding a process. In particular, since the future state is a random variable, there is no guarantee that any prechosen control element p will be a member of the next control set. But even if this difficulty is ignored a policy does not allow opportunity for taking advantage of the vicissitudes of fate, i.e., adapting the policy to benefit from the way a trajectory "happens to come out." In our statistical model, guidance of the process is done by a strategy S(x, t) which is a function of state as well as decision time. It is this dependency on state which provides the mechanism of adaptation. One may conclude from these remarks that feedback (knowledge of current state) is superfluous for deterministic systems, but useful in a statistical framework.

The modifications discussed above are stated formally:

DEFINITION. *A statistical law of motion* is a family $\mathscr{F}$ of n-dimensional* c.d.f.'s conditioned by the variable (x, p, t), where $x \in \Omega$, $t \in A$, and $p \in P(x, t)$.

DEFINITION. *A generalized loss function* L is a real-valued function defined on the triple $(\bar{x}_t, \bar{p}_t, t)$, where $t \in A$, and $\bar{x}_t$ and $\bar{p}_t$ are, respectively, a trajectory and a policy whose domains are restricted to times not less than t.

After the fashion of Chapter 2, when we write $L(\bar{x}, \bar{p}, t)$ we mean $\bar{x}$ and $\bar{p}$ restricted to times not less than t. For brevity, $\mathscr{F}$ will be called simply the law of motion and L a loss function if it is clear that the process under discussion is probabilistic.

DEFINITION. *A strategy* S is a mapping defined on $\Omega \times A$ such that $S(x, t) \in P(x, t)$.

DEFINITION. An *adaptive control process* (abbreviated ACP) is a quintuple $(\Omega, A, P, \mathscr{F}, L)$, where Ω is a state-space, A a set of decision times, P a control set function, $\mathscr{F}$ a statistical law of motion, and L a generalized loss function. If A is finite, the ACP is *truncated.*

A strategy guides the process motion in this sense: If the system is in state-time (x, t), then, under strategy S, the element S(x, t) is selected as the next control. It is to be realized that, unlike the deterministic case, an initial state and choice of strategy do not determine a loss. The loss is a r.v. since L has been made to depend on the trajectory and policy, which are both sequences of r.v.'s prior to motion actually taking place. However, the choice of a

---

* n being the dimension of state space.

strategy and specification of initial state-time do uniquely determine an *expected loss* (denoted E[L(x, S, t)]), as we shall now see.

LEMMA 3.1.   *The trajectory beyond state-time* (x, t) *of a truncated ACP under strategy* S *is a stochastic process whose probability distribution may be found from the process sets.*

This lemma is interesting in itself in addition to being crucial to showing E[L(x, S, t)] is uniquely determined.

PROOF.   Let $\bar{x}_t' = \{x(i)\}_{i>t}$ denote a trajectory beyond decision time t. With this notation, the lemma is implied by the statement "$F(\bar{x}_t' \mid x(t), S, t)$ may be found from $\mathscr{F}$." Induction on m, the number of decision times that remain at time t, establishes this result. If m = 1, then $t = \bar{\bar{A}} = N$ and $\bar{x}_t' = \{x(N + 1)\}$,

$$F(x(N + 1) \mid x(N), S, N) = F(x(N + 1) \mid x(N), S((x, N), N), N) \in \mathscr{F}$$

Suppose the lemma is true for all m < q, and $\xi = N - q$. From probability theory we have

$$F(\bar{x}_\xi' \mid x(\xi), S, \xi) = \int_D F(\bar{x}_{\xi+1}' \mid y, S, \xi + 1) \, dF(y \mid x(\xi), S, \xi), \quad (3.1)$$

where $D = \{y : y \in \Omega$ and for $1 \leqslant i \leqslant n$, i-th component (y) < i-th component $(x(\xi + 1))\}$. Since $N - (\xi + 1) = q - 1$, by the induction hypothesis

$$F(\bar{x}_{\xi+1}' \mid y, S, \xi + 1)$$

may be determined from elements of $\mathscr{F}$. Also,

$$F(y \mid x(\xi), S, \xi) = F(y \mid x(\xi), S(x(\xi), \xi), \xi)$$

is an element of $\mathscr{F}$. Thus (3.1) and the principle of finite induction implies

$$F(\bar{x}_t' \mid x(t), S, t)$$

may be determined from the law of motion in any truncated ACP.

THEOREM 3.1.   *The trajectory beyond state-time* (x, t) *of any ACP under strategy* S *is a stochastic process whose probability distribution is uniquely determined by the process sets.*

Theorem 3.1 differs from the lemma in that the truncation restriction is dropped, but "may be found from" is replaced by "is uniquely determined by," since generally there will be no effective method for computing the values of the distribution function exactly.

PROOF. The theorem follows from Lemma 3.1 and a theorem by Kolmogorov (see Loève, [21, p. 364]) which implies that, if the domain T of the sample function $\bar{x}(t)$ is countable, then the probability law describing the process is uniquely determined by the distributions $F(x(t_1), \ldots, x(t_n))$ for all finite subsets $\{t_1, \ldots, t_n\}$ of T. In our case, T = A is countable and $F(x(t_1), \ldots, x(t_n))$ may be determined from $\mathscr{F}$ as seen in Lemma 3.1. Therefore, the Kolmogorov theorem implies Theorem 3.1, and in fact, from the proof of the Kolmogorov theorem,

$$F(\bar{x}_t' \,|\, x, S, t) = \lim_{n \to \infty} F\left(\{\bar{x}(i)\}_{i=t+1}^n \,\bigg|\, x, S, t\right). \tag{3.2}$$

COROLLARY. *If $g(\bar{x})$ is any function of the trajectory of an ACP, then $E[g(\overline{X}) \,|\, x, S, t]$ is a uniquely determined quantity (which may be found from the process sets if the ACP is truncated). In particular, $E[L(x, S, t)]$ is uniquely determined.*

The corollary follows from Lemma 3.1, Theorem 3.1, and the agreement that we made at the outset of this chapter that all integrands are assumed integrable.

For the same reasons as in the deterministic case, the state-space is a convenient medium for expressing system objectives. Thus we retain the idea of conditions on state-space. However, the property that strategies depend on state allows us to suppress mentioning conditions on state-space by considering only strategies that satisfy these conditions with probability 1. With this restriction, a solution to an ACP problem is any strategy which minimizes the *expected* loss.

DEFINITION. The *set of strategies* associated with an ACP problem are the set of strategies $\mathscr{S}$ such that $S \in \mathscr{S}$ implies

$$P[y \in Q(t+1) \,|\, x, S, t] = 1^* \qquad \text{for all } x \in Q(t), \qquad t \in A, \tag{3.3}$$

$\{Q(i)\}$ being the conditions on state-space as described in the preceding chapter.

---

* By $P[y \in B \,|\, x, S, t]$ we mean the probability that the state at time $t + 1$ will be a member of $B \subset \Omega$, given the state-time is (x, t) and control p is applied. Hereafter we will denote this probability by $P[B \,|\, x, S, t]$. Of course $P[B \,|\, x, S, t] = \int_B dF(z \,|\, x, S, t)$, and, as mentioned in the proof of Lemma 3.1, $F(z \,|\, x, S, t)$ is a member of $\mathscr{F}$.

DEFINITION.   A strategy S is a *solution* to an ACP problem if $S \in \mathscr{S}$ and if (x, t) is the initial state-time of the problem, then for any $S' \in \mathscr{S}$

$$E[L(x, S, t)] \leqslant E[L(x, S', t)]. \tag{3.4}$$

The minimum expected loss criterion, although very popular, is not dictated by logic, and other desiderata may be considered. In particular, dynamic programming finds use in systems featuring the minimax criterion (see Bellman [1, Chapter 10]).

A loss function of particular significance in ACP theory is the subject of the next definition. With this loss function restriction, we will be able to make statements about ACP's that are similar to the dynamic programming and principle of optimality theorems of control process theory.

DEFINITION.   A generalized loss function is *separable* if it admits the representation:

$$L(\bar{x}, \bar{p}, t) = \sum_{i \geqslant t} L(\bar{x}(i), \bar{x}(i + 1), \bar{p}(i), i). \tag{3.5}$$

DEFINITION.   An ACP is *Markovian* if it has a separable loss function and if the trajectory has the Markov property with respect to states and policies, that is, if, denoting any policy and trajectory history of the system by $\bar{p}_0$ and $\bar{x}_0$, respectively, and using the notation of Lemma 3.1,

$$F(\bar{x}'_t \mid \bar{x}_0, \bar{p}_0, t) = F(\bar{x}'_t \mid \bar{x}_0(t), \bar{p}_0(t), t), \tag{3.6}$$

$\bar{x}_0(t)$ and $\bar{p}_0(t)$ being the state and control at time t.

EXAMPLE 3.1.   You are allowed N plays of a fair roulette wheel. At each play t, the number z_t is arrived at in one of two ways: (a) you call for the wheel to be spun, in which case z_t is the resulting number, or (b) you select z_t to be the maximum of $\{z_i : i < t\}$. If you select action (b), the wheel is not spun at the t-th play. The amount you receive for this game (and wish to maximize) is $\sum_{t=1}^{N} z_t$. "00" is worth 0.*

MODEL.   The set of decision times is the initial segment of the natural numbers whose largest element is N. The state-space may be taken to be pairs $(x', x'') = x$, where x' is the largest number thus far received and x'' is the outcome if (a) was chosen in the previous play and x' otherwise. For all pairs (x, t), $P(x, t) = \{a, b\}$ as defined in the problem description. Also

$$L(\bar{x}(j), \bar{x}(j + 1), \bar{p}(j), j) = -z_j = -x''(j + 1).$$

* Processes of this kind are considered in Mallows and Robbins [2].

The law of motion is most conveniently described by probability mass functions:

$$P(y' \mid x, b, t) = \begin{cases} 1, & \text{if } y' = x', \\ 0, & \text{otherwise.} \end{cases}$$

$$P(y'' \mid x, b, t) = \begin{cases} 1, & \text{if } y'' = x', \\ 0, & \text{otherwise.} \end{cases}$$

$$P(y' \mid x, a, t) = \begin{cases} \frac{1}{38}, & \text{if } x' < y' \leqslant 36, \\ 0, & \text{if } y' < x', \\ (x' + 2)/38, & \text{otherwise.} \end{cases}$$

$$P(y'' \mid x, a, t) = \begin{cases} \frac{1}{38}, & \text{if } y'' \neq 0, \\ \frac{1}{19}, & \text{if } y'' = 0. \end{cases}$$

Exercises in Adaptive Control Process Modeling

(1) A and B are two gold mines served by the same mining equipment. At the beginning of each week, either A or B is chosen to be worked. If A is chosen, and the borer does not break, an amount $c_A g_A$ of gold will be extracted by the end of the week, where g_A is the gold in A at the beginning of the week and c_A is a positive constant less than 1. The probability that the rig will break is p_A, in which case no gold will be mined and the operation will be halted indefinitely. Similar parameters c_B, g_B, and p_B are associated with mine B. One wishes a strategy which maximizes the expected gold output from an N-week operation. Model this as an adaptive control process. Is the expected return lower if only policies (state-independent strategies) are considered? This problem is the subject of Chapter II in Bellman [1].

(2) In problem 1 above, suppose that, if mine A is chosen, the amount of g_A which will be removed by the end of the week (if the equipmetn does not fail) is a random quantity whose distribution function is the uniform law on the interval $[0, g_A]$. The amount mined if B is chosen is similarly determined by the uniform distribution on $[0, g_B]$. Find an adaptive control process model for this new situation and indicate whether strategies are to be preferred to policies.

(3) Assume that a gambling house offers q gambles, where the i-th gamble is described by a function $F_i(x; r)$ on R^2 such that for fixed r, F_i is a c.d.f. The interpretation is that, if the gambler wagers r_0 on the gamble i, his next fortune is the r.v. associated with the distribution $F_i(x; r_0)$. For some reason, the gambler needs to change his fortune from \$100 to \$1000 by the end of

N gambling periods. Thus he wishes to find a strategy that maximizes his probability of attaining this end. Model this problem as an ACP problem. The sort of gambling situation described in this problem is the subject of great analytic scrutiny in Dubins and Savage, *How to Gamble If You Must*, McGraw-Hill, New York (1963).

(4) We have N boxes $\{b_i\}_{i=1}^N$ and one of them contains something we are looking for. The i-th box, $1 \leqslant i \leqslant N$, takes t_i minutes to search, where t_i is an integer. The *a priori* probability that box b_i contains the object of our search is p_i. We desire a search procedure that maximizes the probability that the object will be found within M minutes. Model this situation as an ACP problem. Is there any advantage to strategies over policies in this problem? If the loss in not finding the object is c, indicate how one can calculate the worth of an extra minute of search [1, p. 90].

(5) You own 100 shares of a stock which must be disposed of within N days. The market price changes once a day, the next market price being a r.v. whose c.d.f. $F(y;x)$ depends on the current market price x.

(a) Model as an ACP problem the problem of deciding at each day whether or not to sell, so that the expected selling price, under a solution, will be maximized.

(b) Suppose tomorrow's market price depends not only on today's price but also yesterday's. Find a state variable so that the ACP is Markovian.

(6) A detective is allowed to interrogate a group of suspects, one of whom must have committed the murder, and, in the detective's reckoning, all of whom are equally suspect. At each instant in a sequence of times the detective is allowed to ask any of the suspects either of the following questions:

(a) Did you commit the murder?

(b) Has the last question been answered truthfully?

It is estimated that all suspects know who the murderer is, and that suspect i will answer (a) correctly with probability p_i if he is guilty, and with probability r_i otherwise, and he will answer (b) correctly with probability v_i in any event. The problem is to find a strategy so that the detective will be able to maximize the probability that he can pick the guilty party after asking N questions. Model this problem as an ACP problem [1, p. 109].

2. DYNAMIC PROGRAMMING FOR ADAPTIVE CONTROL PROCESSES

The statements we proved about the structure and construction of solutions for control process problems have close counterparts in ACP theory. The separability restriction on loss functions for Markovian ACP's makes the dynamic programming theorem relatively easy to demonstrate. But in the probabilistic scheme, the principle of optimality is a more subtle statement and this author has not been able to prove it for other than truncated systems. In contrast to the development of control process theory, it is more convenient here to begin with the dynamic programming theorem.

LEMMA 3.2. *With a Markovian ACP,*

$$E[L(x, S, t)] = \int (L(x, y, S(x, t), t) + E[L(y, S, t + 1)]) \, dF(y \mid x, S, t). \quad (3.7)$$

PROOF. By the Markovian assumption, denoting $\bar{x}(t) = x$, $\bar{x}(t + 1) = y$,

$$L(\bar{x}, \bar{p}, t) = L(x, y, \bar{p}(t), t) + L(\bar{x}, \bar{p}, t + 1).$$

From probability theory, in the notation of Lemma 3.1,

$$dF(\bar{x}'_t \mid x, S, t) = dF(\bar{x}'_{t+1} \mid y, S, t + 1) \, dF(y \mid x, S, t).$$

$$E[L(x, S, t)] = \int L(\bar{x}, \{S(x(i), i)\}_{i \geqslant t}, t) \, dF(\bar{x}'_t \mid x, S, t)$$

$$= \iint ((L(x, y, S(x, t), t)) \qquad (3.8)$$

$$+ L(\bar{x}'_t, \{S(x(i), i)\}_{i > t}, t + 1))$$

$$\times \, dF(\bar{x}'_{t+1} \mid y, S, t + 1) \, dF(y \mid x, S, t)$$

Let $y \circ \bar{x}'_{t+1}$ denote a trajectory beginning at state-time $(y, t + 1)$. Then

$$E[L(x, S, t)] = \int \left(L(x, y, S(x, t), t) \right.$$

$$+ \left[\int L(y \circ \bar{x}'_{t+1}, \{S(x(i), i)\}_{i > t}, t + 1) \right.$$

$$\left. \times \, dF(\bar{x}'_{t+1} \mid y, S, t + 1) \right] \bigg) \, dF(y \mid x, S, t).$$

The lemma follows, noticing that the quantity enclosed in brackets is identically $E[L(y, S, t + 1)]$.

THEOREM 3.2. (Dynamic Programming for Adaptive Control Processes): *Assume a Markovian ACP problem having initial state-time* (x, j) *has a solution S to all modified problems at time* $j + 1$. *Define*

(1) $0 = \{p : p \in P(x, j)$ *and* $P[Q(j + 1) \mid x, p, j] = 1\}$,

(2) $H(p) = \int (L(x, y, p, j) + E[L(y, S, j + 1)]) \, dF(y \mid x, p, j)$ *for* $p \in 0$.

If there is an element $p^* \in 0$ *such that* $H(p^*) = \min\limits_{p \in 0} H(p)$ *and if*

$$S^*(z, t) = \begin{cases} p^*, & \textit{if } (z, t) = (x, j), \\ S(z, t), & \textit{otherwise} \end{cases}$$

then S^* *is a solution to the problem.*

PROOF.   Let $S' \in \mathscr{S}$. For all $y \in Q(j + 1)$,

$$E[L(y, S', j + 1)] \geqslant E[L(y, S, j + 1)],$$

since S is a solution to all modified problems at time $j + 1$. Since $S' \in \mathscr{S}$, $S'(x, j) \in 0$. Using the lemma, we have

$$\underline{E[L(x, S^*, j)]} \leqslant H\big(S'(x, j)\big)$$

$$\leqslant \int \big(L(x, y, S'(x, j), j) + E[L(y, S', j + 1)]\big)\, dF(y \mid x, S'(x, j), j)$$

$$= \underline{E[L(x, S', j)]}. \tag{3.9}$$

This inequality completes the proof, since (3.9) holds for all $S' \in \mathscr{S}$.*

The DP algorithm for ACP's is much the same as for control processes (CP's). Furthermore, as in the deterministic case, DP is effective both for computation and analysis. Although strategies are functions of state and time whereas a policy depends on time alone, the DP algorithm for ACP's differs little from the CP case. The number of computations and the computer memory requirements for performing DP does not depend on whether the control process is adaptive or not. Specific details of this computational similarity follow the presentation of DP for ACP's. Analytically, DP becomes even more interesting in ACP theory because of the paucity of alternative approaches. For further discussion of this latter topic, see Bellman [3]. For the most part, the remaining chapters of this book are devoted to exploration of the analytic features of DP for ACP's, although future examples will illustrate DP computations.

ALGORITHM.   (*Dynamic Programming for Truncated Markovian Adaptive Control Process.*)

We will inductively (induction being on the number of decision times remaining) construct a strategy S_t which is a solution at all modified problems at time t. If the initial state-time of an ACP problem is (x, j), then S_j is a

* Observe that if S' is a solution to the problem, then $H(S'(x, j)) = \min\limits_{p \in 0} H(p)$.

solution to the original problem. A discussion of the logic of inductive construction of functions may be found in Kelley [4, p. 21].

Let $N = \bar{\bar{A}}$ and, for any $t \in A$, $x \in Q(t)$, let

$$0(x, t) = \{p : p \in P(x, t), \quad \text{and} \quad P[Q(t + 1) \mid x, p, t] = 1\},$$

and

$$H(x, p, N) = \int L(x, y, p, N) \, dF(y \mid x, p, N).$$

If $p^*(x, N)$ is any element of $0(x, N)$ such that

$$H\big(x, p^*(x, N), N\big) = \min_{p \in 0(x,N)} H(x, p, N),$$

then by construction $S_N(x, N) = p^*(x, N)$ must be a solution to all modified problems at time N. Suppose S_{m+1} has been constructed so that it is a solution at all modified problems at time $m + 1$ and

$$H(x, p, m) = \int \big(L(x, y, p, m) + E[L(y, S_{m+1}, m + 1)]\big) \, dF(y \mid x, p, m)$$

and $p^*(x, m)$ is any element of $0(x, m)$ such that

$$H\big(x, p^*(x, m), m\big) = \min_{p \in 0(x,m)} H(x, p, m).$$

If

$$S_m(z, t) = \begin{cases} p^*(x, m), & \text{if } (z, t) = (x, m), \\ S_{m+1}(z, t), & \text{otherwise,} \end{cases}$$

then S_m must be a solution to the modified problems at time m, by Theorem 3.2.

The successful operation of the algorithm depends on the existence of the indicated minima. Section 3 gives a sufficiency condition for this.

EXAMPLE 3.2.   This game is a numerically simplified version of Example 3.1. Instead of a roulette wheel, we have a die. At each decision time t, $t = 1, 2, 3$, you are to choose to have z_t be either (a) the outcome of a toss of a fair die or (b) the highest number thus far thrown. Your object is to maximize $\sum_{t=1}^{3} z_t$.

SOLUTION.   In the manner of Example 3.1, we adopt the following model: $\Omega = \{(x', x'')\}$, where x' is the highest number thus far received, and x'' is the outcome if "a" was the last control, or x' otherwise. The initial state is $(0, 0)$. $A = \{1, 2, 3\}$. $0(x, t) = P(x, t) = \{a, b\}$ for all state-time pairs.

$$L(x(t), x(t + 1), p, t) = -x''(t + 1).$$

The law of motion can be construed easily from Example 3.1. For instance, $P(x''(t + 1) \mid x(t), a, t) = \frac{1}{6}$ for $x''(t + 1) = 1, 2, \ldots$, or 6. Some reflection will reveal that there is nothing to be gained by having decisions depend on x''. The state includes this information solely for the purpose of calculating the loss. Thus our strategies will be defined on (x', t).

For $t = 3$, in the notation of the DP algorithm,

$$H(x', p, 3) = \sum_y L(x, y, p, 3)p(y \mid x', p, 3)$$

$$= -\sum_{y''} y''p(y'' \mid x', p, 3)$$

$$= \begin{cases} -x', & \text{if } p = b, \\ -(\frac{1}{6})(1 + 2 + \cdots + 6) = -\frac{7}{2}, & \text{if } p = a. \end{cases}$$

Then $S_3(x', 3) = p^* \in \{a, b\}$ such that

$$H(x', p^*, 3) = \min \{H(x', a, 3), H(x', b, 3)\}.$$

In Table 3.1, the elements of the solution S_3 to all modified problems occurring at time 3 are displayed.

Table 3.1. Solution and Associated Losses at Modified Problems at Time 3

State coordinate x'	Solution control element $S_3(x', 3)$	Loss $E[L(x, S_3, 3)]$
1	a	$-\frac{7}{2}$
2	a	$-\frac{7}{2}$
3	a	$-\frac{7}{2}$
4	b	-4
5	b	-5
6	b	-6

At time $t = 2$, from the DP algorithm, we have

$$H(x', p, 2) = \sum_y (L(x', y, p, 2) + E[L(y, S_3, 3)])P(y \mid x', p, 2).$$

Therefore,

$$H(x', b, 2) = -x' + E[L(x', S_3, 3)]$$

and

$$H(x', a, 2) = \sum_{y''} -y''p(y'' \mid x', a, 2) + \sum_{y'} E[L(y', S_3, 3)P(y' \mid x', a, 2)].$$

In particular, at $x' = 3$, $t = 2$, our equations are

$$H(3, b, 2) = -(3 + \tfrac{7}{2}) = -\tfrac{13}{2}$$

and

$$H(3, a, 2) = -\tfrac{7}{2} + -[(\tfrac{7}{2})\tfrac{1}{2} + 4(\tfrac{1}{6}) + 5(\tfrac{1}{6}) + 6(\tfrac{1}{6})] = -\tfrac{31}{4}.$$

Since $H(3, a, 2) = -\tfrac{31}{4} < -\tfrac{13}{2} = H(3, b, 2)$, $S_2(3, 2) = p^* = a$. Proceeding in this fashion for all numbers x', we construct Table 3.2,

Table 3.2 Solution and Associated Losses of Modified Problems at Time 2

State coordinate x'	Solution control element $S_2(x', 2)$	Loss $E[L(x', S_2, 2)]$
1	a	$-\tfrac{31}{4}$
2	a	$-\tfrac{31}{4}$
3	a	$-\tfrac{31}{4}$
4	a or b	-8
5	b	-10
6	b	-12

At time 1, obviously the control element should be a. Thus

$$E[L(0, S_1, 1)] = \sum_{y''} - y''P(y'' \mid 0, a, 1) + \sum_{y'} E[L(y', S_2, 2)]P(y' \mid 0, a, 1)$$

$$= -\tfrac{7}{2} - \tfrac{1}{6}(\tfrac{31}{4} + \tfrac{31}{4} + \tfrac{31}{4} + 8 + 10 + 12)$$

$$= -\tfrac{7}{2} - \tfrac{1}{6}(\tfrac{93}{4} + 20) \simeq - 10.7.$$

The negative of $E[L(0, S, 1)]$ is the value of the die game, and a "sensible" gambler should be willing to pay this amount, but no more, for the privilege of playing the game. This last statement overlooks serious difficulties in modeling human behavior which are considered in utility theory [22].

We close this section with some remarks on the importance of DP in ACP theory and requirements on computer memory in executing the DP algorithm. The claims made about the efficiency of DP in control process theory apply to ACP theory even more forcefully. If

$$\overline{\overline{0(x, i)}} = \overline{\overline{0(i)}},$$

independent of x, then as in the control process case, the number of comparisons required to construct a solution is bounded by

$$\overline{\overline{\Omega}} \sum_{i \in A} \overline{\overline{0(i)}}.$$

There are $\overline{\overline{R}}^{\overline{\overline{D}}}$ mappings of a set D into a set R. This implies that, under the above assumptions, there are

$$\prod_{i \in A} 0(i)^{\overline{\overline{\Omega}}}$$

strategies to be compared if a total search for a solution in made. In the statistical case, then, the computational economy of DP is even more impressive than in the deterministic case, where the policy search required only

$$\prod_{i \in A} \overline{\overline{0(i)}}$$

comparisons.

If the solution of an ACP problem having initial time k is not brought out of the computer until calculations are finished, then the memory must be able to accommodate the function S_k, which has $\bar{\Omega}(N - k)$ ordered pairs. (The number $\bar{\Omega}(N - k)$ was obtained by recognizing that the domain of S_k is $\Omega \times \{k, k + 1, \ldots, N\}$.) Consideration of the DP algorithm for ACP's reveals that, to compute S_m, it is sufficient to know only the values

$$E[L(y, S_{m+1}, m + 1)]$$

for all states y. In view of this, by printing out the values $S_m(x, m)$ as they are computed, and storing only the values $E[L(x, S_m, m)]$, the memory needs are reduced to $2\bar{\Omega}$ pairs.

In control process DP, generally the preceding simplication is not valid (unless the loss function happens to be separable), and at each stage of computation, m, the solution policy for each modified problem must be stored in memory. Every such solution is a sequence of $N - m$ control elements, and each state has its own sequence. Therefore DP in the control process case demands storage of $\bar{\Omega}(N - m)$ control elements, which is essentially the same as ACP DP without readout. Roughly speaking, the computational advantage of DP in probabilistic systems is even more striking than in deterministic systems, while the computational bother remains the same.

In this author's opinion, it is in its analytic possibilities in ACP theory that DP displays its greatest strength. Dreyfus [5, p. xiii] has aptly written:

In Chapter 7, the concepts of stochastic and adaptive optimization problems are introduced and the applicability of dynamic programming to this new and challenging area of application is shown. It is the existence of this vast, important class of problems, and the apparent inapplicability of classical methods to these problems, that has motivated much of the research on the dynamic programming method. I feel that it is in this exciting problem domain that the insights and interpretations afforded by dynamic programming, and developed in this book, will ultimately prove most useful.

Bellman, who, as we remarked earlier, coined the phrase "dynamic programming," was the first to exploit it [6, Chapter IX to the end] as a method for extending the intriguing new field of statistics—statistical decision theory—originated by Abraham Wald [7]. The remaining chapters of this book are best classified as contributions to the extension of decision theory by DP ideas. Therefore, we will defer examples of analysis of ACP's until Chapter 4. The balance of this chapter is devoted to an exposition of the current status of ACP theory.

Exercises in Dynamic Programming for Adaptive Control Processes

(1) A truck driver wishes to cover a distance of D miles within N minutes. If he drives at speed r, the probability is p(r) that he will be stopped by the police within a minute. If this happens, the entire minute is wasted, and q additional minutes are lost while a ticket is written. By ACP analysis, indicate formally how to find a strategy for adjusting speed at the beginning of each minute so that the probability of traveling the D miles within N minutes is maximized. Is the speed readjusted at times other than when tickets are given [1, p. 48]?

(2) Consider the caterer's problem, Problem 1 of the exercises in Section 1 of Chapter 2. Assume now that the number of customers on the i-th night is a r.v. whose law is Poisson with parameter λ_i. For simplicity, assume that the rental option is no longer available. Also, there is a penalty Q for each customer who has no napkin. Model this problem as an ACP problem and indicate how to construct a solution.

(3) Assume, in the investor's problem (Problem 7, Section 1, Chapter 2) that the mapping Q is a randomly chosen function which at each day takes on value Q_A with probability p, and Q_B otherwise, where $Q_A(x) \geqslant x$ and $Q_B(x) \geqslant x$ for all x. Indicate formally how to find an investment strategy yielding the greatest possible utility over the N days.

(4) For some interesting assignment of parameters, use a computer to find a solution to Problem 3 above, and also compute the best policy (i.e., an investment schedule which depends on initial but not current capital). Compare the expected utilities of the solution and the optimal policy.

(5) A man attaches utility U to waiting out a queue, but there is a cost c attached to waiting a time unit. The probability that the person at the head

of the line will be finished during unit time is p. Find an ACP model for the problem of deciding whether the man should bother waiting in line, and indicate how to find a solution. In this problem, are strategies to be preferred to policies [1, p. 111]?

(6) A gambler has the opportunity to place money sequentially at 1 to 1 odds on the outcomes of N independent sporting events. He receives advance information on the outcomes of these events, but the probability p is less than 1 that this information is correct. If the gambler initially has a capital of C, how should he bet, at each time j, to maximize his expected capital at the end of N wagers [1, p. 140f].

(7) Let us suppose the gambler in the situation of Problem 6 wishes to maximize the expected value of the logarithm of this final capital. Show that a solution to this problem is to bet at each stage j the amount $qC(j)$, where q is independent of j, the number of the sporting event, and $C(j)$, the gambler's capital prior to making the j-th wager ([1, p. 140f.]. See also Kelley, A new interpretation of information rate, *Bell System Tech. J. 35*, 917–926 (1956), and Bellman and Kalaba, On the role of dynamic programming in statistical communication theory, *IEEE Trans. Information Theory IT-3*, 197–203 (1957).)

(8) Consider the inventory problem, Problem 5, Section 1, Chapter 2, with the following modification: It is no longer known in advance how many items will be required daily. It is known, however, that $N(j)$, the number of items required on the j-th day, $1 \leqslant j \leqslant N$, is the r.v. associated with the Poisson distribution having parameter λ. A cost nK is incurred if the number of items demanded on any day exceeds the number in stock by n. The problem again is to find an optimal ordering strategy.

3. THE EXISTENCE OF SOLUTIONS

The operation of the DP algorithm-depends on the existence of

$$\min_{p \in 0(x,t)} H(x, p, t) = \int (L(x, y, p, t) + E[L(y, S_{t+1}, t + 1)]) \, dF(y \mid x, p, t).$$

$$(3.10)$$

The phase "the existence of the minimum" means that there is some element $p^* \in 0(x, t)$ such that

$$H(x, p^*, t) = \inf_{p \in 0(x,t)} H(x, p, t).$$

The theorem to follow gives a sufficiency condition that is similar to the condition given for the existence of solution to control process problems. Slightly more analysis is required to establish the condition in the ACP context.

THEOREM 3.3 (Existence of Solutions): *Assume a truncated Markovian ACP problem has been specified, wherein* Q(t) *is compact* for each* t, *and such that:*
  (1) *The law of motion is determined by* f(y, | x, p, t), *a real-valued, continuous function which is a probability density for fixed* (x, p, t).
  (2) L(x, y, p, t) *is continuous.*
  (3) 0(x, j) (*as defined in the DP Algorithm*) *equals the compact set* 0(j), *independent of* x, *at all decision times* j.
Then solutions exist at all modified problems, and the dynamic programming algorithm is effective.

PROOF. We prove the theorem by demonstrating that H(x, p, t) (as defined in (3.10), where S_{t+1} is as in the dynamic programming algorithm) is a continuous function. For, if H(x, p, t) is a continuous function of (x, p, t), then it is a continuous function of p alone, x and t remaining fixed [4, p. 102]. If H(x, p, t) is a continuous function of p, then it attains its minimum at some point p* in the compact set 0(t), as a continuous function on a compact set attains its minimum value at some point in that set [4, p. 161].

Induction (on the number of decision times remaining) serves to establish that H is continuous. Let $N = \overline{\overline{A}}$. Then if $p \in 0(N)$

$$H(x, p, N) = \int L(x, y, p, N)f(y \mid x, p, N)\, dy$$

$$= \int_{Q(N+1)} L(x, y, p, N)f(y \mid x, p, N)\, dy, \qquad (3.11)$$

the last equality being a consequence of the definition of 0(N). In Appendix A it is proved that, if $g(x_1, x_2)$ is a continuous function and C a compact set, then the function

$$u(x_1) = \int_C g(x_1, x_2)\, dx_2$$

is continuous. Associating (x, p) with x_1, y with x_2, and observing that products of continuous functions are continuous [8, p. 68], we establish the continuity of H at time N.

----

* Every infinite subset contains an accumulation point.

Now suppose $H(x, p, t)$ is a continuous function for $t = m + 1$, and for each $y \in Q(m + 1)$, let $p^*(y)$ be a control in $0(m + 1)$ such that

$$H(y, p^*(y), m + 1) = \min_{p \in 0(m+1)} H(y, p, m + 1). \quad (3.12)$$

With this notation, we have

$$H(x, p, m) = \int_{Q(m+1)} (L(x, y, p, m) + H(y, p^*(y), m + 1)) f y(\cdot \mid x, p, m) \, dy. \quad (3.13)$$

Sums of continuous functions are continuous [8, p. 68]. Therefore the theorem follows from the statement of Appendix A and the principle of induction if we can show $H(y, p^*(y), m + 1)$ is a continuous function, given $H(y, p, m + 1)$ is continuous. Toward this end, assume $\{y_i\}$ is any sequence converging to y in $Q(m + 1)$, and let p^* be any accumulation point of $\{p^*(y_i)\}$. p^* must be an element of $0(m + 1)$, since for each i, $p^*(y_i) \in 0(m + 1)$ and $0(m + 1)$, being compact, is closed. We have

$$H(y, p^*, m + 1) \geqslant H(y, p^*(y), m + 1) = \min_{p \in 0(m+1)} H(y, p, m + 1). \quad (3.14)$$

Let $p^*(y_{n_i})$ be a subsequence of $p^*(y_i)$ which converges to p^*. For each y_n,

$$H(y_{n_i}, p^*(y), m + 1) \geqslant H(y_{n_i}, p^*(y_{n_i}), m + 1).$$

Thus

$$\lim_{i \to \infty} H(y_{n_i}, p^*(y), m + 1) = H(y, p^*(y), m + 1)$$

$$\geqslant \lim_{i \to \infty} H(y_{n_i}, p^*(y_{n_i}), m + 1)$$

$$= H(y, p^*, m + 1). \quad (3.15)$$

Since p^* was an arbitrary accumulation point, from (3.14) and (3.15) we see that, for any sequence $\{y_i\}$ converging to y,

$$\lim_{i \to \infty} H(y_i, p^*(y_i), m + 1) = H(y, p^*(y), m + 1),$$

which is a definition of continuity.

4. THE PRINCIPLE OF OPTIMALITY

A rather puzzling situation is that, in the engineering literature, DP is used freely, and for the most part correctly, to obtain solutions to statistical problems. When authors justify their procedures, it is usually by appealing to the principle of optimality (see Chapter 2, p. 13), which is often copied verbatim. Such an exposition is odd in two respects: First, the principle of optimality should not be stated axiomatically, since the ACP problem already has sufficient structure to define a solution. The principle of optimality must be proved to be consistent with the criterion already established. This author is unaware of such a published proof and finds it difficult to supply. Second, the principle of optimality is not exactly what is required to justify DP. The principle tells us that any solution to a problem must have the property that it is also a solution to all modified problems which occur. This does not imply that a strategy constructed by DP to have this property is necessarily a solution. That a strategy so constructed is a solution is the statement of the DP theorem for ACP's, which was relatively easy to prove. In our analysis, we have fully justified the use of DP without reference to the principle of optimality. The principle of optimality is of theoretical interest, and it is demonstrated to be true below, under the severe restriction that the Markovian ACP be truncated. This author has found no way to demonstrate even this weak statement without using DP in the proof. For ACP's then, not only have we found the principle of optimality unnecessary to justify DP, but we have to resort to the DP Theorem to justify the principle of optimality. All this is in contrast to our development of these statements for control processes. There we proved the principle of optimality in full generality first and used this principle in proving the DP Theorem.

THEOREM 3.4. (Principle of Optimality for ACP's): *Let* S *be a solution to a truncated Markovian adaptive control process problem having solutions to all modified problems. Then with probability 1,* S *is also a solution to every modified problem which occurs in an observation of the trajectory under* S.

PROOF. Assume $\overline{\overline{A}} = N$, and $x(j)$ is the initial state. The expression "$P[(B, v) \,|\, x(j), S, j]$" will denote the probability that at time v the system is in some state of $B \subset \Omega$, given that $(x(j), j)$ is the initial state-time and that strategy S is followed. We know from Theorem 3.1 that $F(\bar{x}'_j \,|\, x(j), S, j)$ is well defined and may be determined from the process sets. $F(y(v) \,|\, x(j), S, j)$ is a marginal distribution of the distribution above, and thus

$$P[(B, v) \,|\, x(j), S, j] = \int_B dF(y(v) \,|\, x(j), S, j) \qquad (3.16)$$

is also a well-defined quantity for any ACP.

Suppose, contrary to the theorem statement, that there is some set $B \subset \Omega$ which occurs with positive probability at time v under strategy S and such that S is dominated at the modified problem at (y, v) if $y \in B$, i.e., if $y \in B$

there is a strategy S_y such that

$$E[L(y, S, v)] > E[L(y, S_y, v)].\qquad(3.17)$$

$$B_n \equiv \{y: y \in B \text{ and } E[L(y, S, v)] - E[L(y, S_y, v)] > 1/n\}.\qquad(3.18)$$

From the monotone convergence principle of probability measures,

$$\lim_{n \to \infty} P[(B_n, v) \mid x(j), S, j] = P[(B, v) \mid x(j) S, j] > 0,$$

and thus there is a natural number M such that

$$P[(B_M, v) \mid x(j), S, j] > 0.$$

Let S_v be constructed by the dynamic programming algorithm. The theorem assumption that all modified problems have solutions assures that the required minima exist and thus a function S_t as inductively defined in the DP algorithm must exist for $j \leqslant t \leqslant N$. A strategy S' is constructed as follows

$$S'(x, t') = \begin{cases} S_v(x, t), & \text{if } t \geqslant v, \\ S(x, t), & \text{if } t < v. \end{cases}\qquad(3.19)$$

Then

$$E[L(x, S, j)] - E[L(x, S', j)]$$

$$= \sum_{i=j}^{v-1} E[L(x(i), x(i + 1), S(x(i), i), i) \mid x(j), S, j]$$

$$- \sum_{i=j}^{v-1} E[L(x(i), x(i + 1), S'(x(i), i), i) \mid x(j), S', j]$$

$$+ \int_{Q(v)} E[L(x(v), S, v)] \, dF(x(v) \mid x(j), S, j)$$

$$- \int_{Q(v)} E[L(x(v), S', v)] \, dF(x(v) \mid x(j), S', j).\qquad(3.20)$$

Because $S' = S$ if $t < v$,

$$E[L(x(i), x(i + 1), S(x(i), i), i) \mid x(j), S, j]$$

$$= E[L(x(i), x(i + 1), S'(x(i), i)) \mid x(j), S', j]\qquad(3.21)$$

for $j \leqslant i < v$. For the same reason,

$$F(x(v) \mid x(j), S, j) = F(x(v) \mid x(j), S', j).\qquad(3.22)$$

This means

$$E[L(x(j), S, j)] - E[L(x(j), S', j)]$$

$$= \int_{Q(v)} (E[L(x(v), S, v)] - E[L(x(v), S', v)]) \, dF(x(v) \mid x(j), S, j).\qquad(3.23)$$

S' has been defined so that it agrees with S_v for all times equal to or greater than v. Thus S' must be a solution to any modified problem occurring at time v because S_v has that property. So we have

$$E[L(x(v), S, v)] - E[L(x(v), S', v)] \geqslant 0 \qquad \text{if } x(v) \in \Omega - B_M \quad (3.24)$$

and, from the definition of B_M,

$$E[L(x(v), S, v)] - E[L(x(v), S', v)] > 1/M \qquad \text{if } x(v) \in B_M. \quad (3.25)$$

In summary,

$$E[L(x(j), S, j)] - E[L(x(j), S', j)]$$

$$= \int_\Omega (E[L(x(v), S, v)] - E[L(x(v), S', v)]) \, dF(x(v) \,|\, x(j), S, j)$$

$$\geqslant \int_{B_M} (E[L(x, S, v)] - E[(x, S', v)]) \, dF(x(v) \,|\, x(j), S, j)$$

$$> (1/M)P[(B_M, v) \,|\, x(j), S, j] > 0. \qquad (3.26)$$

That $E[L(x(v), S, j)] - E[L(x(v), S', j)] > 0$ contradicts the theorem hypothesis that S is a solution.

5. HISTORY AND SUMMARY OF ADAPTIVE CONTROL PROCESS THEORY

Chapter 4 is devoted to the analysis of some ACP's having unknown statistical parameters, and so we will delay our discussion of ACP's involving parametric learning. Apparently Bellman recognized the applicability of DP to probabilistic systems almost as soon as he thought of DP, for in *Dynamic Programming* [1, p. 80] he mentions obtaining a solution to an ACP in 1950. Several ACP's are discussed in [1], and [6] devotes considerable attention to providing background in probability and game theories so that the scope of ACP's may be more fully appreciated.

Some mathematical statisticians (we mentioned S. Karlin in Chapter 2) have become very interested in ACP's. Particularly noteworthy are the contributions by the game theorist and statistician David Blackwell. His paper "Memoryless Strategies in Finite-State Dynamic Programming" [9] implies that, with truncated ACP's, randomized strategies (we have not considered these) do not give a lower loss than pure strategies.

A stationary strategy is a strategy that is not time-dependent, i.e., $S(x, t) = S(x, 1)$ for all $t \in A$. Blackwell's paper "On the Functional Equation of

Dynamic Programming" [10] gives a sufficient condition that an ACP have a stationary solution. The papers by Blackwell [11–13] and by Derman [14] are powerful mathematical explorations proving the existence of stationary solutions for certain ACP's in which motion does not depend on time and loss at each stage is determined by a time-invariant function multiplied by a discount factor.

Following a research path opened by Shapley [15], Denardo [16] has revealed a unified theory for CP's and ACP's which enjoy a "contraction property." Processes within the scope of Denardo's theory include the discounted ACP's of Blackwell and Derman referenced above.

A very readable book that thoroughly examines a finite-state ACP having time-independent motion and discounted loss is Ronald Howard's *Dynamic Programming and Markov Processes* [17]. Another fundamental study of a finite-state ACP has been made by Eaton and Zadeh [18]. Dubins and Savage's *How to Gamble If You Must* [19] is a mathematically profound book treating an interesting problem in gambling which is much like an ACP by methods related to DP.

It is possible to bring DP to bear on ACP's by viewing them as having a structure very similar to that of control processes, and this is the way Denardo [20] proceeds. We describe his analysis with our notation. The law of motion is deterministic, but the elements in the range are not states, but subsets of state-space. One guides a process with a strategy. Suppose a solution S_{m+1} has been constructed to all modified problems at time $m + 1$ and for any $p \in 0(x, m)$,

$$H_p(z, t) = \begin{cases} p, & \text{if } (z, t) = (x, m), \\ S_{m+1}(z, t), & \text{otherwise.} \end{cases}$$

Then, provided $E[L(x, S, t)]$ has Mitten's monotonicity property (see [8] of Chapter 2), if

$$E[L(x, H_{p^*}, m)] = \min_{p \in 0(x,m)} E[L(x, H_p, m)], \tag{3.27}$$

H_{p^*} is a solution to the modified problem at (x, m). Thus if

$$S_m(x, t) = \begin{cases} p^*(x), & \text{if } t = m, \\ S_{m+1}(x, t) & \text{otherwise} \end{cases}$$

where x is any state in $Q(m)$ and $p^*(x)$ is found as in (3.27), S_m is a solution of all modified problems at time m.

In Denardo's method, L and $\mathscr{F}$ do not appear in the process description, but only terms $E[L(x, S, t)]$. In our studies of the sequential design of experiments, the nature of the statistical law of motion $\mathscr{F}$ is of the greatest significance, and thus we have decided not to repress this set.

Readers acquainted with the literature of dynamic programming will recognize that the process we have been calling an adaptive control process is usually termed a "stochastic process," the designation "adaptive control process" being reserved for processes having unknown statistical parameters. We justify the break with tradition by asserting that processes with unknown statistical parameters are conveniently modeled as ACP's (our terminology). Chapter 4 explains in detail how this modeling is to be done. It is really feedback, the dependency of strategy on state, that is at the heart of probabilistic processes, whether or not the parameters are known. Since feedback makes adaptation possible, this author finds it more natural to think of any process whose control functions are state-dependent as an adaptive process.

REFERENCES

1. Bellman, R. (1957). *Dynamic Programming*. Princeton University Press, Princeton, N.J.
2. Mallows, C. and H. Robbins (1964). Some problems of optimum sampling strategy. *J. Math. Anal. Appl. 8*, 90–103.
3. Bellman, R. (1962). On the foundations of a theory of stochastic variational processes. *Proc. Symp. Appl. Math. 13*, 275–286.
4. Kelley, J. L. (1955). *General Topology*. Van Nostrand, Princeton, N.J.
5. Dreyfus, S. E. (1965). *Dynamic Programming and the Calculus of Variations*. Academic Press, New York.
6. Bellman, R. (1961). *Adaptive Control Processes*. Princeton University Press, Princeton N.J.
7. Wald, A. (1947). *Sequential Analysis*. Wiley, New York.
8. Apostol, T. M. (1957). *Mathematical Analysis*. Addison-Wesley, Reading, Mass.
9. Blackwell, D. (1964). Memoryless strategies in finite-state dynamic programming. *Ann. Math. Statist. 35*, 863–865.
10. Blackwell, D. (1961). On the functional equation of dynamic programming. *J. Math. Anal. Appl. 2*, 273–276.
11. Blackwell, D. (1962). Discrete dynamic programming, *Ann. Math. Statist. 33*, 719–726.
12. Blackwell, D. (1965). Discounted dynamic programming. *Ann. Math. Statist. 36*, 226–235.
13. Blackwell, D. (1967). Positive dynamic programming. *Fifth Berkeley Symposium on Mathematical Statistics and Probability*, Vol. I, pp. 415–418. University of California Press, Berkeley.
14. Derman, C. (1965). Markovian sequential control processes—denumerable state space. *J. Math. Anal. Appl. 10*, 295–302.
15. Shapley, L. S. (1953). Stochastic games. *Proc. Nat. Acad. Sci. U.S.A. 39*, 1095–1100.
16. Denardo, E. (1967). Contraction mappings in the theory underlying dynamic programming. *SIAM Rev. 9*, 165–177.
17. Howard, R. A. (1960). *Dynamic Programming and Markov Processes*. M.I.T. Press, Cambridge, Mass.
18. Eaton, J., and L. Zadeh (1962). Optimal pursuit strategies in discrete-state probabilistic systems. *Trans. ASME Ser. D. J. Basic Eng. 84*, 23–29.

19. Dubins, L. E., and L. J. Savage (1963). *How to Gamble If You Must.* McGraw-Hill, New York.
20. Denardo, E. (1965). *Sequential Decision Processes.* Dissertation, Northwestern University. University Microfilms No. 65-12,069.
21. Loève, M. (1963). *Probability Theory,* 3rd ed., Van Nostrand, Princeton, N.J.
22. Luce, R. D., and H. Raiffa (1957). *Games and Decisions.* Wiley, New York.

Chapter 4

ADAPTIVE CONTROL PROCESSES IN TWO-ARMED BANDIT THEORY

I. AN ELEMENTARY TWO-ARMED BANDIT PROBLEM AND ITS SOLUTION

A type of problem arising in the sequential design of statistical experiments which has been energetically studied in the statistical literature is often denoted the "two-armed bandit (TAB) problem." In TAB problems as in sequential design problems generally, a statistical parameter is to be estimated while an adaptive control process develops in time. One seeks a strategy in which expenses incurred in estimation and control are balanced to give an overall minimum expected loss. TAB problems successfully capture the flavor of the very interesting statistical questions and viewpoints that arise in the study of sequential design problems.

An elementary and interesting example of a TAB problem [1, pp. 215–216] is the following: Two slot machines, I and II, are presented to the designer. Machine I pays a dollar with probability r (independently of previous plays and machine II) and II pays a dollar with probability s. The designer is allowed N free plays which may be distributed between the two machines as he wishes. He is told the value of s, but not the value of r. However, he is told that r is selected by a random experiment described by the c.d.f. F(r). The designer wishes to play the machines so as to maximize his take during the N plays.

If the designer knew r, the solution would be obvious: For all plays choose I or II according to whether r is greater or less than s. If the events "r is greater than s" and "r is less than s" both have positive probability, finding the best method of play is a more subtle task. By identifying this TAB problem as a truncated Markovian ACP, the theory of Chapter 3 leads rather directly to solutions.

In formulating the TAB as an ACP, the set A of decision times is $\{1, 2, \ldots, N\}$. It is convenient to take as state space

$$\Omega = \{(m, n, v): m, n, v \text{ are non-negative integers and } m + n + v \leqslant N\}.$$

The components of the state (m, n, v) indicate I has achieved m successes

49

(payoffs) and n failures and II has a history of v successes. For all state-time pairs, the control set is {I, II}, the elements of which specify the choice of machine for the next play. In the context of bandits, it is more natural to speak of gain G than loss L. The object of the problem then becomes to find a strategy that maximizes the expected gain. A gain function $G(\bar{x}, \bar{p}, j)$ which counts the total payoffs is

$$G(\bar{x}, \bar{p}, j) = \sum_{i \geqslant j} g(\bar{x}(i), \bar{x}(i + 1)), \tag{4.1}$$

where $g(x, y) = m_2 - m_1 + v_2 - v_1$ if $x = (m_1, n_1, v_1)$ and $y = (m_2, n_2, v_2)$. Thus maximizing the expected value of G is equivalent to maximizing the expected return from the game. Expression (4.1) makes it plain that G is separable.

Finally we derive the law of motion from the process sets. Assume the system is in state (m, n, v) when II is played. Since it is known that II pays with probability s,

$$P[(m, n, v + 1) \mid (m, n, v), II] = s,$$
$$P[(m, n, v) \mid (m, n, v), II] = 1 - s, \tag{4.2}$$
$$P[\text{Any other state} \mid (m, n, v), II] = 0.$$

If machine I is played,

$$P[(m + 1, n, v) \mid (m, n, v), I]$$

$$= \int P[(m + 1, n, v) \mid (m, n, v), I, r] \, dF(r \mid m, n)$$

$$= \int r \, dF(r \mid m, n), \tag{4.3}$$

where the assumption that machine I and II are independent has allowed us to write $dF(r \mid m, n)$ instead of $dF(r \mid m, n, v)$. To determine $F(r \mid m, n)$ from the process sets, it is helpful to consider the r.v. (R, m, n) whose sample space is $[0, 1] \times J \times J$, J being the non-negative integers. As plays on I are Bernoulli trials with parameter r,

$$P[m, n \mid r] = \binom{m + n}{m} r^m (1 - r)^n.$$

By the "total probability rule" of elementary probability theory,

$$P[B \times \{m\} \times \{n\}] = \int_B P[m, n \mid r] \, dF(r)$$

$$= \binom{m + n}{m} \int_B r^m (1 - r)^n \, dF(r). \tag{4.4}$$

If U and V are any subsets of $[0, 1] \times J \times J$ such that $P[V] > 0$, from the definition of conditional probability

$$P[U \mid V] = P[U \cap V]/P[V].$$

Let $U = [0, r] \times \{m\} \times \{n\}$ and $V = [0, 1] \times \{m\} \times \{n\}$. Since

$$F(r \mid m, n) = P[[0, r] \mid (m, n)],$$

$$F(r \mid m, n) = P[[0, r] \times \{m\} \times \{n\}]/P[[0, 1] \times \{m\} \times \{n\}], \quad (4.5)$$

which from (4.4) gives

$$F(r \mid m, n) = (P[(m, n)])^{-1} \int_0^r (r')^m (1 - r')^n \, dF(r'), \quad (4.6)$$

where

$$P[(m, n)] = \int_0^1 (r')^m (1 - r')^n \, dF(r').$$

From (4.6)

$$dF(r \mid m, n) = (P[(m, n)])^{-1} r^m (1 - r)^n \, dF(r).$$

Recalling (4.3), in summary we have

$$P[(m + 1, n, v) \mid (m, n, v), I] = (P[(m, n)])^{-1} \int_0^1 r^{m+1}(1 - r)^n \, dF(r) \equiv r(m, n).$$

Similarly it is seen that

$$\quad (4.7)$$

$$P[(m, n + 1, v) \mid (m, n, v), I] = 1 - r(m, n),$$

and $P[\text{Any other state} \mid (m, n, v), I] = 0$.

In summary, the elements of the law of motion are

$$P((m, n, v + 1) \mid (m, n, v), II) = s,$$

$$P((m, n, v) \mid (m, n, v), II) = 1 - s$$

$$P((m + 1, n, v) \mid (m, n, v), I) = r(m, n), \quad (4.8)$$

$$P((m, n + 1, v) \mid (m, n, v), I) = 1 - r(m, n)$$

$$P[\text{All other states} \mid (m, n, v)] = 0.$$

This completes the identification of TAB components as ACP sets. We have yet to show that, under this identification, the ACP is Markovian. We have noted that the loss function is separable. Therefore, it remains only to demonstrate that the law of motion has the Markovian property with respect to states, i.e.,

$$P(x(t) \mid p, x(t - 1), \ldots, x(1)) = P(x(t) \mid p, x(t - 1)). \quad (4.9)$$

If $p = $ II, because s is known, nothing is to be gained from statistical analysis; we already have complete statistical information of machine II. Thus (4.9) holds. If $p = $ I, the validity of (4.9) is established by showing

$$F(r \mid I, x(t-1), \ldots, x(1)) = F(r \mid I, x(t-1)). \qquad (4.10)$$

For if (4.10), then

$$\underline{P(x(t) \mid I, x(t-1), \ldots, x(1))} = \int_0^1 r \, dF(r \mid I, x(t-1), \ldots, x(1))$$

$$= \int_0^1 r \, dF(r \mid I, x(t-1)) = \underline{P(x(t) \mid I, x(t-1))}$$

$$(4.11)$$

if $x(t)$ and $x(t-1)$ are, respectively, $(m + 1, n, v)$ and (m, n, v). A similar relationship exists if $x(t)$ is $(m, n + 1, v)$. The event $(x(t-1), \ldots, x(1))$ is a particular ordering of successes and failures among the two machines. For the analysis to follow, it is notationally helpful to consider the equivalent event $\{(a, b, k)\}_{k=1}^{t-1}$, where a is I or II according to which machine was played at time k, and b is 0 or 1, the payoff resulting from that play. Then

$$P(x(t-1), \ldots, x(1) \mid I, r) = P[\{(a, b, k)\}_{k=1}^{t-1} \mid I, r]. \qquad (4.12)$$

We now use the hypothesis that the machines are independent to express the above probability as the product of probabilities of the trials on each machine separately. Thus if T_j denotes the times at which machine j is played, $j = $ I, II, then

$$P[x(t-1), \ldots, x(1) \mid I, r] =$$

$$= P[\{(a, b, k)\}_{k \in T_I} \mid I, r] P[\{(a, b, k)\}_{k \in T_{II}} \mid I].$$

$$(4.13)$$

We have deleted r from the dependency in the second term as the Bernoulli parameter s of machine II is unrelated to r.

Assume $x(t-1) = (m, n, v)$. Then there have been m successes in $m + n$ trials on machine I. Thus

$$P[\{(a, b, k)\}_{k \in T_I} \mid I, r] = r^m (1-r)^n. \qquad (4.14)$$

The reasoning that led to (4.6) also justifies

$$F(r \mid I, x(t-1), \ldots, x(1)) = \frac{\displaystyle\int_0^r P(x(t-1), \ldots, x(1) \mid I, r') \, dF(r')}{\displaystyle\int_0^1 P(x(t-1), \ldots, x(1) \mid I, r') \, dF(r')}. \qquad (4.15)$$

Using our new notation and the decomposition in (4.13),

$$\underline{F(r \mid I, x(t-1), \ldots, x(1))}$$

$$= \frac{P[\{(a, b, k)\}_{k \in T_{II}} \mid I] \int_0^r P[\{(a, b, k)\}_{k \in T_I} \mid I, r'] \, dF(r')}{P[\{(a, b, k)\}_{k \in T_{II}} \mid I] \int_0^1 P[\{(a, b, k)\}_{k \in T_I} \mid I, r'] \, dF(r')}$$

$$= \frac{\int_0^r (r')^m (1-r')^n \, dF(r')}{\int_0^1 (r')^m (1-r')^n \, dF(r')} = \underline{F(r \mid I, x(t-1))}. \qquad (4.16)$$

The last equality is a recollection of (4.6). Thus we have established the Markovian property of the law of motion with respect to state.

There are other plausible ways of modeling this TAB as an ACP. The choice of state variable was motivated by the wish to have a variable of minimal dimension which is a sufficient statistic for r and also provides enough information (the total number of successes) to calculate the gain.

As we have established a truncated Markovian ACP model for the TAB, we may now routinely employ the DP algorithm to obtain a solution. Substituting "G" (gain) for "L" (loss) and therefore "max" for "min" in the DP algorithm,

$$S_N(x, N) = p^*(x, N),$$

where $p^*(x, N) \in \{I, II\}$ and

$$E[G(x, y, p^*(x, N), N)] = \max_{p \in \{I, II\}} \int G(x, y, p, N) \, dF(y \mid x, p, N).$$

Using (4.1),

$$\int G(x, y, p, N) \, dF(y \mid x, p, N) = \int g(x, y) \, dF(y \mid x, p, N)$$

$$= \begin{cases} r(m, n), & \text{if } p = I \text{ and } x = (m, n, v), \\ s, & \text{if } p = II. \end{cases}$$

Therefore,

$$E[G((m, n, v), S_N, N)] = \max \{r(m, n), s\},$$

where $r(m, n)$ is as defined in (4.7).

And for the variable k, $1 \leqslant k < N$, by substitution of the appropriate terms into the dynamic programming algorithm of Chapter 3, we have

$$E[G((m, n, v), S_k, k)] = \max \begin{cases} r(m, n)\big(1 + E[G((m + 1, n, v), \\ \quad S_{k+1}, k + 1)]\big) \\ + \big(1 - r(m, n)\big)E[G((m, n + 1, v), \\ \quad S_{k+1}, k + 1)] \\ = H((m, n, v), I, k), \\ s\big(1 + E[G((m, n, v + 1), S_{k+1}, k + 1)]\big) \\ + (1 - s)E[G((m, n, v), S_{k+1}, k + 1)] \\ = H((m, n, v), II, k). \end{cases} \quad (4.17)$$

One may interpret the expressions in the set on which "max" operates thusly: If I is played, with probability $r(m, n)$, a payoff is received and the system "moves" to state $(m + 1, n, v)$. If a solution S_{k+1} is followed from this point on, a return $E[G((m + 1, n, v), S_{k+1}, k + 1)]$ may be expected. Alternatively, with probability $1 - r(m, n)$, there is no payoff and the system "moves" to state $(m, n + 1, v)$. If a solution is followed for the rest of the game, the expectation will be $E[G((m, n + 1, v), S_{k+1}, k + 1)]$. The terms associated with playing II lend a similar interpretation. The quantities S_{k+1}, $E[G(x, S_{k+1}, k + 1)]$ have presumably been computed in the previous step of the algorithm. In (4.17), $H(x, I, k)$ may be regarded as the expected value of the game if I is played next. Among the details emerging from these computations is the value of the entire game, $E[G((0, 0, 0), S_1, 1)]$. This is the greatest amount the designer should be willing to pay for the privilege of playing the game. The game is worth less than this amount to a player who does not use a solution.

In regard to solving a TAB problem in detail, two lemmas concerning the structure of solutions are presented. These lemmas, in conjunction with dynamic programming, reduce the computational complexity to a minimum.

LEMMA 4.1. *For the TAB problem of Section 4.1, a solution* S *depending only on the first two components of the state variable may be found by dynamic programming. Consequently, whenever* x *and* x' *are states which agree on the first two components,*

$$E[G(x, S, j)] = E[G(x', S, j)].$$

PROOF. The lemma is demonstrated by induction on $b = N - j + 1$, the number of decision times that remain at state-time (x, j). Referring to the analysis preceding (4.17), if $b = 1$,

$$E[G((m, n, v), S_N, N)] = \max \{r(m, n), s\}.$$

As the term to the right does not depend on v, it is immediate that $S(x', N) = p$ is a solution at (x', N) if p gives a maximum expected gain at (x, N) and x and x' differ only in v. Therefore,

$$E[G((m, n, v), S_N, N)] = E[G((m, n, v'), S_N, N)]$$

for all non-negative v'.

Suppose for $b < \xi$ the lemma is true. From (4.17),

$$E[G((m, n, v), S_{N-\xi+1}, N - \xi + 1)]$$
$$= \max \{H((m, n, v), I, N - \xi + 1), H((m, n, v), II, N - \xi + 1)\}, \quad (4.18)$$

where

$$H((m, n, v), I, N - \xi + 1)$$
$$= r(m, n)(1 + E[G((m + 1, n, v), S_{N-\xi+2}, N - \xi + 2)])$$
$$+ (1 - r(m, n))E[G((m, n + 1, v), S_{N-\xi+2}, N - \xi + 2)] \quad (4.19)$$

and

$$H((m, n, v), II, N - \xi + 1) = s(1 + E[G((m, n, v + 1), S_{N-\xi+2}, N - \xi + 2)])$$
$$+ (1 - s)E[G((m, n, v), S_{N-\xi+2}, N - \xi + 2)]$$
$$(4.20)$$

At time $N - \xi + 2$, only $\xi - 1$ decision times remain and thus the inductive hypothesis implies the terms

$$E[G((q_1, q_2, q_3), S_{N-\xi+2}, N - \xi + 2)]$$

are not dependent on q_3. Hence from examination of (4.19) and (4.20) it is apparent that $H((m, n, v), p, N - \xi + 1)$ does not depend on v for $p = I, II$, and consequently in view of (4.18) neither does a DP solution nor the value

$$E[G((m, n, v), S_{N-\xi+1}, N - \xi + 1)].$$

This completes the inductive proof.

LEMMA 4.2.   *If* S *is a solution to the TAB problem of Section 4.1 and for some state-time pair* (x, t) *occurring with positive probability,* $S(x, t) = II$, *then* $E[G(x, S, t)] = (N - t + 1)s$.

The proof of this lemma constitutes Appendix B.

Lemma 4.1 is interesting, but its greatest significance is that it is needed in the proof of Lemma 4.2. S_{II} denotes the strategy which maps all state-time pairs into II. It is evident that

$$E[G(x, S_{II}, t)] = (N - t + 1)s$$

for all pairs (x, t), and that

COROLLARY. *Under the conditions of Lemma 4.2, S_{II} is a solution to the modified problem at (x, t).*

The corollary is equivalent to the statement that, once a solution calls for playing II, once may as well play II for the rest of the game and that our TAB problem may be viewed as the stopping rule problem, "When is it no longer feasible to experiment with machine I?" The corollary is assurance that the set of strategies that map (m, n, v) into II if $v \neq 0$ is essentially complete. That is, the value of the game is not diminished if only these strategies are allowed. Consequently, only states with $v = 0$ are considered. Further, the analysis in the example to follow takes advantage of the corollary in considering only those strategies which never turn to I after II has been played.

EXAMPLE 4.1.   The preceding TAB analysis is illustrated in a numerical example displayed in detail in Table 4.1. Figure 1 graphically displays the solution. The parameters of this example are $N = 6$, $s = 0.6$, and $dF(r)/dr$ the uniform density on the unit interval. As mentioned in the paragraph above, only states with $v = 0$ need be considered. For brevity, $(m, n, 0)$ is written (m, n) in the example. For $dF(r)/dr$, the uniform density, (4.6) implies

$$\frac{dF(r \mid (m, n))}{dr} = \frac{r^m(1 - r)^n}{\displaystyle\int_0^1 r^m(1 - r)^n \, dr}.$$

Thus

$$r(m, n) = \frac{\displaystyle\int_0^1 r^{m+1}(1 - r)^n \, dr}{\displaystyle\int_0^1 r^m(1 - r)^n \, dr} = \frac{m + 1}{m + n + 2}.$$

The computer program used to calculate the preceding solution is listed in Appendix C, and it may easily be modified to find the solution to longer, more sophisticated games.

If the designer were told the value of r as well as s at the beginning of each game (and therefore played machine I or II according to whether or not r

Table 4.1. Equations, Values and Strategies for a TAB Problem

Parameters: $N = 6$, $s = 0.6$, and $dF(r)/dr$ is the uniform density on $[0, 1]$. $K(x, t) \equiv E[G(x, S_t, t)]$.

Equations	State x	Value $E[G(x, S_j, j)]$	Solution $S_j(x, j)$
j = 6			
	0, 5	0.60	II
	1, 4	0.60	II
$K(x, 6) = \max \begin{cases} (m + 1)/7 = H(x, I, 6) \\ 0.6 = H(x, II, 6) \end{cases}$	2, 3	0.60	II
	3, 2	0.60	II
	4, 1	0.72	I
	5, 0	0.86	I
j = 5			
	0, 4	1.20	II
$K(x, 5) = \max \begin{cases} ((m + 1)/6)(1 + K((m + 1, n), 6)) \\ + (1 - (m + 1)/6)K((m, n + 1), 6) \\ = H(x, I, 5) \\ 2(0.6) = H(x, II, 5) \end{cases}$	1, 3	1.20	II
	2, 2	1.20	II
	3, 1	1.34	I
	4, 0	1.67	I
j = 4			
	0, 3	1.80	II
$K(x, 4) = \max \begin{cases} ((m + 1)/5)(1 + K((m + 1, n), 5)) \\ + (1 - (m + 1)/5)K((m, n + 1), 5) \\ = H(x, I, 4) \\ 3(0.6) = H(x, II, 4) \end{cases}$	1, 2	1.80	II
	2, 1	1.88	I
	3, 0	2.41	I
j = 3			
$K(x, 3) = \max \begin{cases} ((m + 1)/4)(1 + k((m + 1, n), 4)) \\ + (1 - (m + 1)/4)K((m, n + 1), 4) \\ = H(x, I, 3) \\ 4(0.6) = H(x, II, 3) \end{cases}$	0, 2	2.4	II
	1, 1	2.4	II
	2, 0	3.02	I
j = 2			
$K(x, 2) = \max \begin{cases} ((m + 1)/3)(1 + K((m + 1, n), 3)) \\ + (1 - (m + 1)/3)K((m, n + 1), 3) \\ = H(x, I, 2) \\ 5(0.6) = H(x, II, 2) \end{cases}$	0, 1	3.0	II
	1, 0	3.48	I
j = 1			
$K((0, 0), 1) = \max \begin{cases} \frac{1}{2}(1 + K((1, 0), 2)) \\ + \frac{1}{2}K((0, 1), 2) = H(x, I, 1) \\ 6(0.6) \qquad = H(x, II, 1) \end{cases}$	0, 0	3.74	I

were greater than s), the expected worth of each play would be $E[\max \{s, R\}]$. In the case where $s = 0.6$ and R is uniformly distributed on the unit interval, as in the example above,

$$E[\max \{s, R\}] = sP(R < s) + P[R \geqslant s]E[R \mid R \geqslant s]$$
$$= 0.6(0.6) + (0.4)(1/0.4)\int_{0.6}^{1} r \, dr = 0.68.$$

One may observe in Table 4.2 that, as N, the length of the game, increases, the average return converges toward this idealized return.

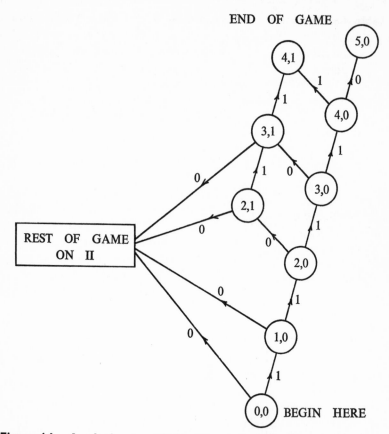

Figure 4.1. A solution to a TAB with one machine known. The node indicates the current state and that machine I is to be played. The branch labeled with the outcome resulting from that play leads to the choice for the next play.

Table 4.2. Average Return per Play for TAB (various N)

N	$E[G((0, 0, 0), S_1, 1)]/N$
6	0.62
10	0.64
25	0.655
60	0.665
100	0.6676
500	0.6755
1000	0.6773

A final remark on this example is that, if a person played machine I, irregardless of history, for 60 plays, his expected reward would be $30. If he played the machines randomly, on the average he would collect $33. If he stuck to machine II, he could expect $36. With a solution, the expected return is $39.90—over 10% higher than the best of these other strategies. The best policy (a strategy not dependent on state) for this example is: Play machine II for the entire game. Thus feedback is worth almost $4.

2. DISCOUNTED AND MULTI-ARMED BANDITS

An impetus for the research reported in this chapter is an intriguing bandit problem proposed [1, p. 215] and deeply studied by Richard Bellman. This problem differs from the problem studied in the preceding section in that the statistician is allowed infinitely many plays. Moreover, the payoff from the i-th play is discounted by a^{i-1}, where $0 < a < 1$; thus the maximum return of the game is bounded by

$$\sum_{i=0}^{\infty} a^i = \frac{1}{1-a}.$$

Bellman [2] has proved that a solution exists to this game, but there are no known methods for finding this solution in finitely many computations. From a practical standpoint, this difficulty is resolved in the analysis to follow.

Given any positive number ε and initial state x, the methods of Section 4.1 may be employed to construct a strategy S' in a finite, predetermined number of computations such that, if S is a solution to the discounted bandit problem, then

$$E[G(x, S, 1)] - E[G(x, S', 1)] < \varepsilon. \tag{4.21}$$

That is, the expected gain using S' differs from the value of the game by a number less than ε. To construct S', choose a number N such that

$$\sum_{i=N+1}^{\infty} a^i < \varepsilon.$$

If a solution S_N can be exhibited to the discounted game truncated after the N-th play, and S' is defined:

$$S'(x, t) = \begin{cases} S_N(x, t), & \text{if } t \leqslant N, \\ \text{I or II picked arbitrarily}, & t > N, \end{cases}$$

then S' satisfies (4.21). This assertion follows from the following equations:

$$\sum_{i=N+1}^{\infty} \left(E[G(x(i), x(i+1), i) \mid x, S, 1] - E[G(x(i), x(i+1), i) \mid x, S', 1] \right)$$

$$\leqslant \sum_{i=N+1}^{\infty} E[G(x(i), x(i+1), i)] \mid x, S, 1] \leqslant \sum_{i=N+1}^{\infty} a^i < \varepsilon,$$

and

$$\sum_{i=1}^{N} E[G(x(i), x(i+1), i)] \mid x, S, 1] - E[G(x(i), x(i+1), i)] \mid x, S', 1] \leqslant 0.$$

The last equation is a restatement of the assumption that S_N is a solution to the truncated problem, having observed that S, restricted to times $\{1, \ldots, N\}$, is a strategy for that problem.

As there are only finitely many strategies for the truncated problem, at least one of them must be a solution. This solution S_N may be found efficiently by modeling the problem as an ACP problem. The identification is the same as the formulation of Section 1, except for an appropriate modification of the loss function to accommodate the discount factor. We now have, in the notation of (4.1),

$$G(\bar{x}, \bar{p}, k) = \sum_{i \geqslant k} g(x(i), x(i+1), i),$$

where, if $x = (m', n', v')$ and $y = (m, n, v)$,

$$g(x, y, n) = a^{n-1}(m - m' + v - v').$$

As this gain function is separable, the discussion of Section 1 suffices to demonstrate that this discounted TAB problem is a truncated ACP problem; the dynamic programming algorithm may therefore be applied routinely.

The investigations of Section 1 admit extension is another way. Instead of having to choose between only two machines, there may be M machines available, where M may even exceed the number of plays in the game. In this multi-armed bandit problem, each machine is statistically independent of the others. The statistician is supplied with c.d.f.'s $F(r, k)$, $1 \leqslant k \leqslant M$. $F(r, k)$ describes the experiment by which the Bernoulli parameter r_k, the payoff probability of the k-th machine, is determined. As before, the statistician seeks to select machines on the basis of payoff history as the process unfolds, so as to maximize his expected gain.

In modeling this bandit problem, the state space is conveniently taken to be sequences of the form $\{(m, n, i)\}_{i=1}^{M}$, where the triple (m, n, k) indicates that machine k has a history of m successes in $m + n$ trials. The control set

$P(x, t)$ for every state-time pair (x, t) is $\{1, 2, \ldots, M\}$ where $k \in P(x, t)$ indicates machine k is to be played. In the spirit of (4.7), define

$$r(m, n, k) = \frac{\int_0^1 r^{m+1}(1 - r)^n \, dF(r, k)}{\int_0^1 r^m(1 - r)^n \, dF(r, k)}$$

The development of the law of motion in Section 1 justifies the relations

$$P[(m + 1, n, k), (m, n, j) \quad j \neq k \mid \{(m, n, i)\}_{i=1}^M, k] = r(m, n, k),$$

$$P[(m, n + 1, k), \quad (m, n, j) \quad j \neq k \mid \{(m, n, i)\}_{i=1}^M, k] = 1 - r(m, n, k),$$

$$P[\text{Any other state} \mid \{(m, n, i)\}_{i=1}^M, k] = 0.$$

The set of c.d.f.'s generated as x and k range over their domains comprises the law of motion for this multi-armed bandit. A demonstration that the law of motion has the Markov property with respect to states follows the earlier argument for two machines. A gain function which counts the number of successes is

$$G(\bar{x}, \bar{p}, k) = \sum_{i=k}^N g(x(i), x(i + 1), i),$$

where

$$g(x, y, n) = \sum_{j=1}^M m_j' - m_j, \quad \text{if } x = \{(m_j, n_j, j)\} \text{ and } y = \{(m_j', n_j', j)\}.$$

The observation that this gain function is separable is all that is now needed to confirm that the multi-armed bandit problem is a truncated ACP problem and may therefore be solved by the DP algorithm.

Figure 4.2 displays a solution to the multi-armed bandit problem with $N = 5$, $M = 2$, and $dF(r, 1)/dr$ and $dF(r, 2)/dr$, both the uniform density on the unit interval. The equations and values associated with this problem are contained in Appendix D. This bandit problem illustrates a general truth of dynamic programming: The values of the problems and modified problems are always unique, but the solution need not be unique. Appendix E contains the numerical values associated with the dynamic programming analysis of this example as well as a computer program for performing the dynamic programming algorithm.

B. Gluss [14] has given a method for finding a solution to this bandit problem, and R. Steck [15] has published a computer procedure for implementing Gluss's method.

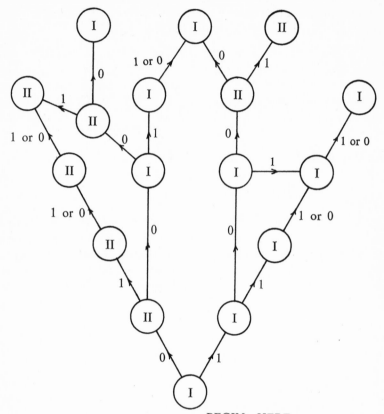

BEGIN   HERE

Figure 4.2.   A solution to a TAB with both machines unknown. The node indicates the machine to be played. The branch labeled with the outcome resulting from that play leads to the node indicating the choice for the next play.

3. CONSTRUCTION OF INITIAL SEGMENTS OF EXACT SOLUTIONS TO THE DISCOUNTED, INFINITE-PLAY BANDIT

In this section, the discounted bandit introduced in Section 2 is studied further. As mentioned before, it is known that solutions exist to this problem, but there are no known techniques for finding them in finitely many computations. In Section 2 we gave a method that yields an approximate solution S' in finitely many computations; that is, given $\varepsilon > 0$ and initial state x, finitely many computations suffice for finding a strategy S' such that if S is a solution,

$$E[G(x, S, 1)] - E[G(x, S', 1)] < \varepsilon.$$

The analysis to follow is an alternative approach to the discounted bandit problem. Briefly, given any positive integer M, by the method to be presented, finitely many computations will yield a strategy S^M which agrees with some exact solution restricted to times not greater than M. This being the case, if S is a solution that coincides with S^M at times not greater than M, it is immediate that

Value (Discounted Game) $- E[G(x, S^M, 1)]$

$$= E[G(x, S, 1)] - E[G(x, S^M, 1)]$$
$$\leqslant E[G(x, S, M)] \leqslant a^{M-1}/(1 - a).$$

Roughly speaking, if the designer follows S^M, any bad decisions, i.e., decisions which jeopardize his total return, occur after time M. For very large M, it may be impossible to solve this TAB problem by the method of Section 2 because of the large memory requirement associated with long games. However, we still may be able to find the first M elements of a solution to the infinite game by the algorithm to be revealed in this section. It is theoretically interesting that it is possible to compute the initial segment of an exact solution, for mathematics is rife with objects that are known to exist but have never been found. The analysis to follow reveals intriguing facets in the logical structure of the discounted bandit not seen in Section 2.

The analysis begins with a theorem by Bellman, which we rephrase in our terminology.

THEOREM 4.1. *For any infinite-play, discounted two-armed bandit problem as in Section 2, there exists a double sequence* $\{s(m, n)\}$ *such that the strategy*

$$[S((m, n, v), t) = S(m, n) = I] \Leftrightarrow [s(m, n) > s]$$

is a solution. Also, $s(m + 1, n) > s(m, n) > s(m, n + 1)$.

Theorem 4.1 is essentially Theorem 2 of [2], which the reader is invited to consult for a proof.

Let $\{u(n)\}$ be the sequence defined by

$$u(n) = \begin{cases} \max (B = \{j : s(j, n) \leqslant s \text{ and } 0 \leqslant j \leqslant n\}), \text{if } B \neq \varnothing, \text{ the empty set,} \\ -1, \quad \text{if } B = \varnothing. \end{cases}$$

Then the strategy

$$(S(m, n) = I) \Leftrightarrow (m > u(n))$$

is the solution described in Theorem 4.1. Thus knowledge of $\{u(n)\}$ is equivalent to possession of a solution. For any positive integer N let G_N and S_N denote the gain function and solution associated with the discounted TAB truncated after N plays and let $u_N(n)$ be the largest non-negative integer such that $j \leqslant n$ and

$$E[G_N((j, n), S_N, 1)] = (1 - a^N)s/(1 - a)* \qquad (4.22)$$

or $u_N(n) = -1$ if no non-negative integer satisfies (4.22). That is, $u_N(n)$ is the largest number j such that playing II at the first play of a truncated game with initial state (j, n) is not bad.

Always,

$$u_N(n) \geqslant u(n), \qquad (4.23)$$

for plays on II cannot change the first two components of state and thus, if S is the solution of Theorem 4.1, once II is played, II is played for the rest of the game and, as $S(u(n), n) = $ II,

$$E[G((u(n), n), S, 1)] = s/(1 - a). \qquad [4.24]$$

Suppose $j > u_N(n)$. Then

$$E[G_N((j, n), S_N, 1)] > (1 - a^N)s/(1 - a).$$

If

$$S'_N(z, t) = \begin{cases} S_N(z, t), & \text{if } t \leqslant N, \\ \text{II}, & \text{otherwise,} \end{cases}$$

then

$$E[G((j, n), S'_N, 1)] = E[G_N((j, n), S_N, 1)] + a^N s/(1 - a) > s/(1 - a),$$

which would contradict (4.24) and the assumption that S is the solution of Theorem 4.1 if it is not also true that $j > u(n)$.

Let S''_N be a best strategy for the truncated game that always plays I at time 1. Suppose

$$E[G_N((u_N(n), n), S_N, 1)] - E[G_N((u_N(n), n), S''_N, 1)] > a^N/(1 - a). \qquad (4.25)$$

Then we have that

$$u_N(n) \leqslant u(n), \qquad (4.26)$$

for, if S'' is any strategy for the infinite game such that $S''((u_N(n), n), 1) = $ I, and $\hat{S}_N$ an arbitrary extension of S_N to decision times $N + 1, \ldots,$

$$E[G((u_N(n), n), S'', 1)] \leq E[G((u_N(n), n), S''_N, 1)] + a^N/(1 - a)$$

$$< E[G_N((u_N(n), n), S_N, 1)]$$

$$\leq E[G((u_N(n), n), \hat{S}_N, 1)].$$

* The corollary of Lemma 4.2 applies to the truncated discounted TAB as well as to the TAB of Section 1.

which implies S'' cannot be a solution. Therefore, if (4.25) holds,

$$u_N(n) = u(n).$$

The justification of our method requires another fact.

THEOREM 4.2.   *Assume* s, *the Bernoulli parameter of machine II, is chosen by a random mechanism whose probability distribution has no mass points. Then with probability 1, for each number* n, *(4.25) holds for some finite number* N.

The proof of this theorem constitutes Appendix F.

ALGORITHM *for Finding an Initial Segment of an Exact Solution for the Discounted, Infinite-Play TAB.*

The algorithm exploits the fact that specification of $\{u(n)\}$ is equivalent to a solution, and therefore the segment $\{u(n)\}_{n=1}^{M}$ determines best plays for the first M decision times. The algorithm shows how $\{u(n)\}_{n=1}^{M}$ may be constructed.

ALGORITHM.   Given any positive integer M, for each n, n = 1, 2, . . . , M,

(1) Pick any positive N and compute $u_N(n)$, S_N, and S''_N. This may be done efficiently by using the dynamic programming algorithm as indicated in Section 4.2.

(2) Calculate

$$q = E[G_N((u_N(n), n), S_N, 1)] - E[G_N((u_N(n), n), S''_N, 1)].$$

(3) If $q > a^N/(1 - a)$, $u(n) = u_N(n)$. If $q \leqslant a^N/(1 - a)$, repeat steps 1 to 3 using a positive integer $N' > N$ in place of N. If s has been chosen by some person or process which has not consciously sought out the one value for which

$$E[G((u(n), n), S, 1)] = E[G((u(n), n), S'', 1)],$$

then, as one who has leanings toward the subjectivist view of probability, in light of Theorem 4.2, I will have confidence that the algorithm will be effective. We have had some computing experience without running into any instance where the algorithm required N to be greater than 50 for a = 0.8 and 0.9. Of course, N will be sensitive to the value of "a." A program for and a readout of an initial segment of a solution for the TAB problem with a = 0.8, s = 0.6, and F(r), the uniform distribution on [0, 1], constitutes Appendix G.

4. IN QUEST OF A GENERAL BANDIT PROBLEM

The scheme of the earlier sections of this chapter has been to convert given TAB problems to ACP's and proceed with analysis through the theory developed in Chapter 3. However, our studies have revealed that bandits have a character of their own: Their laws belong to the multinomial family, their returns at each play are independent of their history and the histories of other bandits in the process, etc. Arising from this character is the property that ACP models of the different bandits have similar state and control spaces and laws of motion. In view of this modeling consistency, in this section the direction of analysis is reversed. Reflection on abstract properties of Markovian ACP's suggests that more general bandit problems can easily be handled by dynamic programming. But also from this point of view we see that certain conditions must prevail in bandit problems in order that dynamic programming be effective.

There is a considerable mechanism available to us in ACP theory that we did not use at all. In particular, we did not have control sets depend on state or time, and the return from each play depended on the control chosen only through the probability behavior. In short, it is transparent that we could levy a charge dependent on the machine chosen, and the time when it is chosen. Also, we can assume the machines available to the designer depend on his state (or equivalently, his current fortune) and the decision time. In fact, there is no end of intriguing complications which do not impair the effectiveness of our dynamic programming approach: The cost of playing a machine may depend on the designer's fortune, there may be machines whose reward is the permission to play certain machines which in turn have incomplete probabilistic description, etc.

On the other hand, our understanding of ACP's helps us delimit the class of bandit problems we can solve. Suppose instead of receiving \$1 for a success at the k-th trial, the designer receives a ticket marked "k." After N plays, if he presents his tickets at a certain booth, he will receive

$$\$\left(\sum_{i=1}^{q} z(i) + \left(\sum_{q+1}^{N} z(i) \right)^2 \right),$$

where $z(i)$ is equal to 1 or 0 according to whether or not a ticket marked "k" has been presented. The use of dynamic programming for this problem, as easy and natural as it would seem, is not justified with the ACP model of this chapter because the gain function is not separable. (If the bandits were deterministic, curiously enough, dynamic programming would be effective, as the above function happens to be type M.)

In Chapter 5, two problems in pattern recognition will be examined that have mathematical description identical to bandits B_1 and B_2. B_1 is a bandit problem in which the designer is allowed to keep only the last payoff. B_2 is any bandit problem which has the added feature that, for each state-time (x, t), the house offers the designer the option of receiving b(x, t) (depending on state-time (x, t)) if he will terminate play. The effect of the offer is to increase the control set by the element p' and offer for consideration the return $G(x, y, p', t) = b(x, t)$.

In place of r(m, n) in the dynamic programming algorithm for TAB's, one can use other estimators, but such procedures are on shaky logical ground and lead to strategies that grate the intuition. For example, suppose one elects to use the maximum likelihood estimator $\hat{r}(m, n) = m/n$ in place of r(m, n). Then in the TAB with s known, as $\hat{r}(0, 1) = 0$, dynamic programming will direct the designer to play machine II for the rest of the game if he does not receive a payoff on the first trial, regardless of the value of s. Other methods of classic statistics also seem poorly suited for analysis of TAB problems.

5. HISTORY AND SUMMARY OF TWO-ARMED BANDIT THEORY

Bellman [2] credits W. R. Thompson, a prominent biostatistician, with being the first to formulate a TAB problem. Thompson [3] phrased the problem in the context of testing the effectiveness of two drugs thusly: On the basis of case histories, which of two drugs of unknown effectiveness should be used on the next patient in order that the maximum number of lives may be saved? Thompson's studies, which were published in 1933, preceded the development of sequential analysis and were tedious and inconclusive.

The statistician Herbert Robbins recognized the significance of the TAB problem as a general question of economical design of statistical experiments. His 1952 paper [4] contains a very eloquent description of the merits of sequential design over fixed sample-size analysis, which even yet dominates statistics curricula. For a TAB problem with both machines unknown and no prior distribution, Robbins gave a strategy in which average payoff at each play converges to the payoff of the best machine as the number of plays becomes large. In a later paper [5], Robbins posed the problem of finding good strategies for the TAB just described among strategies that depend only on the outcomes of the last r plays. In this paper, a strategy is proposed and its expected return calculated. Isbell [6] and Smith and Pyke [7] have continued the investigation, exhibiting successively better strategies.

In 1956, two papers were published which phrased the TAB problem in Bayesian fashion, assuming prior probabilities on the Bernoulli parameters of the unknown machines. From one of these [2], the principal results were incorporated in our Theorem 4.1. The other paper, by Bradt, Johnson, and

Karlin [8], while not exhibiting a general method for solving truncated TAB problems, shows that solutions to such problems have interesting properties. For instance, the strategy that picks the machine with the highest posterior Bayes estimate for the Bernoulli parameter is dominated for some truncation length. Also, the strategy "stick with a winner" is dominated. Finally, it is seen that under the guidance of a solution the expected return may be a decreasing function of decision time. This paper concludes with a very careful analysis of the TAB problem of our Section 1, bringing out some structural aspects not derived in our exposition.

In partial converse to certain results in [8], Feldman [9] proved that, in the special case in which either machine I has a Bernoulli parameter p or (1 − p) and correspondingly II has (1 − p) or p, the strategy that chooses the machine with the higher posterior expectation is a solution. Among the remarks and generalizations in [9] are that, under the guidance of a solution in this TAB problem, the wrong machine will be played only finitely many times in an infinite-play game and that the results hold for more general machines than those having Bernoulli payoffs.

Vogel [10] has performed a study of truncated Bayesian bandits which are specialized in that, until some time 2k, which is to be chosen on the basis of payoff history, the bandits are played alternatively. At time 2k the designer is to choose one of the machines for exclusive play during the remaining decision times of the process. Some asymptotic results are derived for large decision times.

In another paper [11], Vogel initiated a new branch of non-Bayesian bandit studies by considering the case that Bernoulli parameters may change in some restricted fashion from play to play.

TAB theory is related in objective to the theory of sequential analysis, a discipline initiated and strongly pursued by the late Abraham Wald [12], an outstanding mathematical statistician. Sequential analysis is concerned with the following problem: A designer plays a game against nature. That is, he chooses an act "a" from some set $\mathscr{A}$ and nature chooses an act θ from a set $\boldsymbol{\theta}$. The designer experiences the resultant loss $L(a, \theta)$. The assumption is that nature chooses θ randomly, the probability law being a known c.d.f. $F(\theta)$. The "sequential analysis" arises in that, instead of choosing the act "a" directly, the designer has the option of inferring the value θ chosen by nature after observing an outcome, $x(1)$, of an experiment whose c.d.f., $F(x;\theta)$, depends upon θ in a known manner. There is a cost involved in performing the experiment. After $x(1)$ has been observed, the designer again has the option of choosing an act directly or making an additional observation, $x(2)$, and so the decision problem repeats itself *ad infinitum* until the designer stops testing and chooses an act. The object of analysis is to find a stopping rule and a decision rule for choosing the terminal act so that the total expected loss is minimized, relative to $F(\theta)$.

The TAB theory represents a notable generalization of the sequential analysis framework in that more than one experiment (or machine) is available. In contrast to sequential analysis, in the TAB problem we seek a rule for choosing among experiments as well as perhaps a stopping rule. This requires significant modification in mathematical structure of the single-population sequential analysis situation. It will be seen in Chapter 5 that the sequential analysis feature of playing a game at the termination of experimentation can be treated in ACP theory. The mathematical structure for the sequential analysis problem proposed by Wald and adopted by others does not easily admit Markovian ACP modeling because the experimentation cost can depend on the sequence of observations in a completely general fashion.

In their book *Theory of Games and Statistical Decisions* [13] Blackwell and Girshick give a beautiful exposition of the theory of statistical games and sequential analysis. Chapters 9 and 10 of [13] consider problems and techniques close to our TAB analysis—the essential difference being the non-Markovian nature of the former. In particular, with some modification, a method revealed on page 256 of [13] could be made into an effective if not an efficient algorithm for the TAB problems of this chapter.

A merit of TAB problems is that they pose sequential design problems in a way that is readily understood and mechanically not overly tedious to analyze. Without the stimulus and intrigue of TAB's, this author might not have developed an interest in the statistical aspects of dynamic programming.

Exercises on Two-Armed Bandits and Related Learning Processes

(1) In view of the way slot machines actually operate, the following multi-armed bandit is of interest. A machine k has $\psi(k)$ possible money prizes $\{a(k, i)\}_{i=1}^{\psi(k)}$ which occur at each trial of machine k according to the multi-nomial distribution with parameters

$$P[a(k, i)] = r(k, i), 1 \leqslant i \leqslant \psi(k).$$

M such machines are available. The values of the probabilities $r(k, i)$ are not known to the designer, but he does know that they are selected by a random mechanism whose c.d.f. is

$$F(\{r(k, i)\}_{1 \leqslant i \leqslant \psi(k)}), 1 \leqslant k \leqslant M.$$

The designer is allowed N free plays and wishes to maximize his take. Model this problem as an ACP problem and indicate the dynamic programming algorithm for this problem.

(2) Suppose after the fashion of *How to Gamble If You Must* (reference 19 of Chapter 3), the player wishes only to maximize the probability that he will win some fixed sum, V. Indicate how the ACP model for the TAB of Section 1 should be modified to accommodate this desideratum. Indicate the dynamic programming algorithm. Is Lemma 4.2 true under this new circumstance? Does the solution differ from the solution of Section 1?

(3) A bakery which adjusts its price daily can sell bread at either c_1 or c_2 per loaf. If the price is c_j, $j = 1, 2$, the number of customers on that day will be an observation of a r.v. whose law is Poisson with parameter λ_{c_j}. Unfortunately, the bakery does not know what λ_{c_j} is, but in Bayesian fashion has prior densities $f_{\lambda_{c_1}}(x)$, $f_{\lambda_{c_2}}(x)$ describing the unknown parameters.* The bakery wants a strategy for adjusting its price at each day on the basis of customer history so as to maximize its expected income over an N-day operation. Find an ACP model for this problem and indicate how a solution can be constructed. (*Hint:* The sample mean is a sufficient statistic for the Poisson parameter.)

(4) You are offered 2 to 1 odds in coin tossing, where you win whenever a tail occurs. However, you suspect the coin is biased—a density $f(p)$ being your prior Bayes estimate of p, the probability that a head appears. Give an ACP model and solution to the problem of determining at each toss whether or not to continue playing. Assume the length of the game cannot exceed N throws.

(5) Obtain a computer solution to Problem 4 above with $N = 25$ and $f(p)$, the uniform density on the interval $[\frac{1}{2}, 1]$.

(6) Can a solution be found for Problem 4 if the length of the game is not bounded?

(7) Referring to the investment problem, Problem 3, Section 2, Chapter 3, imagine that p is fixed but not known to the designer. Assume, however, that the designer knows there are but two values, p_1 and p_2, to which p can possibly be equal, and *a priori* the probability that $p = p_1$ is v. Give an ACP model for this new situation and indicate how a solution can be constructed.

(8) This problem is a continuation of the stock-disposal problem, Problem 5, Section 1, Chapter 3. Suppose $F(y; x)$ in the problem is not known to the designer. However, he does know that $F(y; x) = F_{r_0}(y; x)$ is the normal distribution with unit variance and mean $x - r_0$. The number r_0 is

---

* L. J. Savage, *The Foundations of Statistics*, Wiley, New York (1954), is a standard reference on the Bayes approach.

not known, but *a priori* it is assumed to have been chosen by the random process whose law is uniform on $[-1, 1]$. The unknown parameter r_0 is assumed fixed throughout the N-day decision period. By modeling this problem as an ACP problem, indicate how a solution may be found.

(9) This problem is a continuation of the "man in a queue" problem, Problem 5, Section 2, Chapter 3. The situation is now that the man does not know the value of p, but assumes it has been chosen by the random process whose distribution is uniform on $[0, 1]$ and is the same for all customers. Find a solution to this new problem and explain why now strategies are preferred to policies, in contrast to the original queuing problem.

REFERENCES

1. Bellman, R. (1961). *Adaptive Control Processes, A Guided Tour.* Princeton University Press, Princeton, N.J.
2. Bellman, R. (1956). A problem in the sequential design of experiments. *Sankhyā 16,* 221–229.
3. Thompson, W. R. (1933). On the likelihood that one unknown probability exceeds another in view of the evidence of two samples. *Biometrika 25,* 285–294.
4. Robbins, H. (1952). Some aspects of the sequential design of experiments. *Bull. Amer. Math. Soc. 58,* 527–535.
5. Robbins, H. (1956). A sequential decision problem with finite memory. *Proc. Nat. Acad. Sci. U.S.A. 42,* 920–923.
6. Isbell, J. R. (1959). On a problem of Robbins. *Ann. Math. Statist. 30,* 606–610.
7. Smith, C. V., and R. Pyke (1965). The Robbins-Isbell two-armed-bandit problem with finite memory. *Ann. Math. Statist. 36,* 1375–1386.
8. Bradt, R. N., S. M. Johnson, and S. Karlin (1956). On sequential designs for maximizing the sum of n observations. *Ann. Math. Statist. 27,* 1060–1074.
9. Feldman, D. (1962). Contributions to the "two-armed bandit" problem. *Ann. Math. Statist. 33,* 847–856.
10. Vogel, W. (1960). A sequential design for the two armed bandit. *Ann. Math. Statist. 31,* 430–443.
11. Vogel, W. (1960). An asymptotic minimax theorem for the two armed bandit problem. *Ann. Math. Statist. 31,* 444–451.
12. Wald, A. (1947). *Sequential Analysis.* Wiley, New York.
13. Blackwell, D., and M. A. Girshick (1954). *Theory of Games and Statistical Decisions.* Wiley, New York.
14. Gluss, B. (1957). A note on a computational approximation to the two machine problem. *Information and Control 1,* 268–275.
15. Steck, R. (1964). A dynamic programming strategy for a two machine problem. *Math. Comp. 18,* 285–291.

Chapter 5

ADAPTIVE CONTROL PROCESSES IN PATTERN RECOGNITION THEORY

I. AN INTRODUCTION TO THE PATTERN RECOGNITION MACHINE

Figure 5.1 illustrates a popular model for the pattern recognition machine (PRM). $W = \{w_i\}_{i=1}^{h}$ is a finite set of patterns. At the k-th stage of operation,

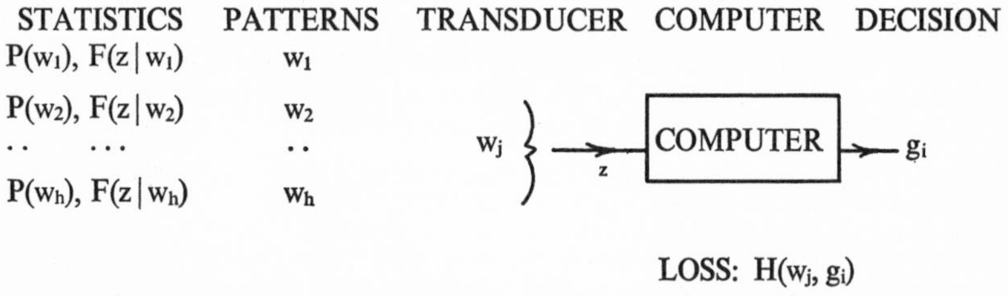

STATISTICS	PATTERNS	TRANSDUCER	COMPUTER	DECISION
$P(w_1), F(z\mid w_1)$	w_1			
$P(w_2), F(z\mid w_2)$	w_2			
.. ...	..	w_j	COMPUTER	g_i
$P(w_h), F(z\mid w_h)$	w_h			

LOSS: $H(w_j, g_i)$

Figure 5.1. A model for a pattern recognition machine.

with probability $p(w_j)$, at the exclusion of the other patterns, pattern w_j is presented to the transducer, which makes a real v-tuple measurement z on it. Prior to obtaining the measurement, Z is a random variable described by the conditional c.d.f. $F(z \mid w_j)$, which may not be known to the system designer. On the basis of the measurement z, the computer attempts to guess the input pattern, w_j. If the guess is denoted by g_i, the loss $H(w_j, g_i)$ is then realized. This is a convenient and accurate statistical model for a device that is to recognize hand-printed letters, distinguish code words sent through a noisy channel (i.e., a detector), or classify radar returns. The generality of this model has been discussed in Chapter 1, Section 2.

If all the statistics are known in advance, the design of an optimum PRM (one in which the expected loss is minimized) is a simple matter: if W' denotes the unknown input pattern (which is a r.v.), then the Bayes risk is achieved by the decision rule [1, Chapter 6]:

$$d^*(z) = w^* \in W, \tag{5.1}$$

where

$$E[H(W', w^*) \mid z] = \min_{g \in W} E[H(W', g) \mid z] \qquad (5.2)$$

and

$$E[H(W', g) \mid z] = \sum_{w \in W} H(w, g) P(w \mid z). \qquad (5.3)$$

If the distributions $F(z \mid w_i)$ are either discrete or absolutely continuous, $P(w \mid z)$ can be found from the known statistics $\{P(w_i), F(z \mid w_i)\}$ by Bayes rule, as we will see in Section 2.

Generally, the statistics are unknown however. To describe this situation, the unknown distribution $F(z \mid w_i)$ is assumed to be a member of the metric space of c.d.f.'s, $\mathscr{F}_i$, which is indexed by the continuous one-to-one function θ_i:

$$\theta_i : R^m \rightarrow \mathscr{F}_i. \qquad (5.4)$$

The image of the real m-tuple r_i will be denoted by $F(z \mid r_i)$. The designer does not know which member of $\mathscr{F}_i$ is $F(z \mid w_i)$. In terms of R^m, parameter space, the parameter θ_i' whose image, $F(z \mid \theta_i')$, is $F(z \mid w_i)$ is unknown. In Bayesian fashion, the parameter θ_i' is supposed chosen by a random experiment whose c.d.f., $F(\theta_i)$, is known to the designer. The designer is never told the outcome of this experiment, but as we shall see observations on the transducer provide the opportunity for statistically inferring this unknown outcome.

Roughly speaking, the more accurately θ_i' is known, in a statistical sense, the lower the Bayes risk will be. For this reason, typically the system designer will seek to improve his estimate of θ_i' by performing experiments on the transducer before putting the PRM into operation. By repeatedly presenting the pattern w_i to the transducer, a sequence $\{Z_j\}$ of independent r.v.'s, each distributed according to $F(z \mid w_i)$, is obtained. From this sequence the experimenter would expect to improve his estimate of θ_i' by some statistical technique. The procedure of repeatedly presenting a known input to the transducer is called "supervised learning" in the communication theory literature. Before proceeding to the central subject of this chapter (which is the use of adaptive control process theory to make the best use of a learning period), some results are presented with the intention of convincing the reader that generally a learning period is worthwhile. The author believes that these statements are stronger than publications on this matter in the communication theory literature because here, for the first time, bounds for error are made for the small-sample case. Previous studies [2, 11] have been concerned only with asymptotic convergence.

To show that all parameters may be learned arbitrarily accurately in finitely many observations, it suffices to show that one parameter may be learned arbitrarily accurately in a finite number of observations; attention here will be restricted to a single pattern, obviating the need for the subscript "i." In the following analysis, it will be assumed that the sample space and the parameter space are both the real numbers. It is further assumed that the metric ρ on $\mathscr{F}$ is:

$$\rho(F_1, F_2) = \sup_{z \in R} |F_1(z) - F_2(z)|. \tag{5.5}$$

Finally, the following definition is needed:

DEFINITION. Let $\{Z_i\}_{i=1}^{n}$ be a sequence of independent r.v.'s each distributed as the same r.v. Z. If $\{z_{\sigma(i)}\}_{i=1}^{n}$ is an order statistic* of an observation on the sequence and the function $F_n(z)$ is constructed thusly:

$$F_n(z) = \begin{cases} 0, & \text{if } z \leqslant z_{\sigma(1)}, \\ (h-1)/n, & \text{if } z_{\sigma(h-1)} < z \leqslant z_{\sigma(h)}, \\ 1, & \text{if } z > z_{\sigma(n)}, \end{cases} \tag{5.6}$$

then $F_n(z)$ is the *empirical distribution function of Z in n trials.*

The fundamental statement of this study is:

THEOREM 5.1. *Let $\{Z_i\}$ be a sequence of independent trials of Z, a r.v. distributed according to some arbitrary c.d.f. F. Let $\{F_i\}$ be the sequence of functions such that F_n is the empirical distribution function constructed from the first n terms of $\{Z_i\}$. Then, given any positive numbers c_1, c_2, a number N, dependent on c_1 and c_2 but not on F, may be found such that*

$$P\left[\sup_{n>N} \rho(F_n, F) > c_1\right] < c_2. \tag{5.7}$$

In fact, if M is a positive integer such that $1/M < c_1/8$, Q_1 is a positive integer such that

$$(c_1/8)^2(c_2/2(M+1)) > \sum_{j=Q_1}^{\infty} 1/j^2$$

and $Q_2 = 16/c_1$, then N will fulfill the properties of the theorem if $N = \max \{Q_1^2, Q_2\}$. The proof of Theorem 1, which constitutes Appendix H, is similar to the argument used by Loève [3, p. 20] in proving the Glivenko-Cantelli theorem. Loève does not, however, make explicit how N can be

---

* An order statistic is a permutation σ of the terms of the observation $\{z_i\}_{i=1}^{n}$ such that $z_{\sigma(i)} > z_{\sigma(k)}$ implies $i > k$.

determined. Formulas for finding N in the special case that $F(z)$ is known to be continuous are readily available for weak-law convergence in the statistics literature (see for instance [4, p. 339f]).

The importance of the theorem to the theory of pattern recognition is made clear by the corollary to follow, which establishes that, with arbitrarily small probability of error, the unknown parameter may be estimated arbitrarily closely in a finite, predetermined number of learning observations.

COROLLARY 1. *If the index function θ*

$$\theta : R \to \mathscr{F} \tag{5.8}$$

is continuous and the prior c.d.f. $F(\theta)$ is supplied, then, given any two positive numbers c_1, c_2, a statistic V and a number N may be found such that

$$P\left[\sup_{n > N} |V(\{z_i\}_{i=1}^n) - r'| > c_1 \right] < c_2, \tag{5.9}$$

if $F(z \mid r')$ is the c.d.f. associated with the r.v. Z and $\{z_i\}_{i=1}^n$ is an observation on n trials of Z.

PROOF. Let M be any number such that

$$\int_{(-M,M)} dF(\theta) > 1 - c_2/2. \tag{5.10}$$

Such an M exists, since

$$\lim_{M \to \infty} \int_{(-M,M)} dF(\theta) = \int_{\lim_{M \to \infty} (-M,M)} dF(\theta) = 1.$$

Since $[-M, M]$ is compact and θ is continuous, $\theta([-M, M])$ is compact.* As θ is 1-1, θ^{-1} exists and must therefore be uniformly continuous on $\theta([-M, M])$.* Let c_1' be any positive number such that if $\rho(F(z/r_1), F(z/r_2)) < c_1'$ and $r_1, r_2 \in [-M, M]$, then $|r_1 - r_2| < c_1$. Let V be any function such that

$$V(\{z_j\}_{j=1}^n) = r_n \in (-M, M)$$

and

$$\rho(\theta(r_n), F_n) = \rho(F(z \mid r_n), F_n(z)) < c_1'/2$$

if such an r_n exists. Otherwise r_n is chosen arbitrarily. By the theorem, if $r' \in (-M, M)$, for all but finitely many n, r' will be a candidate for r_n and hence the above condition will be satisfied. If $\rho(\theta(r_n), F_n) < c_1'/2$, by the triangle property for metrics,

$$\rho(\theta(r_n), F) \le \rho(\theta(r_n), F_n) + \rho(F_n, F) < c_1', \tag{5.11}$$

---

* Demonstration of these facts are to be found in [5, pp. 72, 75].

provided $\rho(F_n, F) < c_1'/2$. From the definition of c_1', $\rho(\theta(r_n), F) < c_1'$ implies $|r_n - r'| < c_1$. Referring to the theorem, if N is the number associated with $(c_1'/2, c_2/2)$, this number suffices for the conclusion of the corollary because the probability that r' is not in $(-M, M)$ is less than $c_2/2$, and if r' is in $(-M, M)$ then the probability is greater than $1 - c_2/2$ that, for all $n > N$, $|r_n - r'| < c_1$.

COROLLARY 2. *If the domain of the index function* θ *is compact, the conclusion of corollary 1 is valid regardless of whether the prior c.d.f. is available.*

PROOF. The function of the prior c.d.f. in Corollary 1 is to ensure the existence of a compact set which occurs with probability greater than $1 - c_2/2$. Here this compactness condition is satisfied to begin with.

In the preceding analysis it has been assumed that the sample space and the parameter space are one-dimensional. No difficulty is encountered if the parameter space has higher dimension, but the multi-dimensional sample space does introduce analytic complications we have not had time to resolve. J. Kiefer has published (*Pacific J. Math.*, **11**, 649–660) a powerful analysis which implies Theorem 5.1 is true in the n-dimensional case provided "may be found" is replaced by "exists."

2. SUPERVISED LEARNING AS AN ADAPTIVE CONTROL PROCESS

Return now to the general situation in which there are h patterns, $w_1, \ldots, w_h$, whose parameters $\theta_1', \ldots, \theta_h'$ are to be "learned." Recall θ_i is a variable mapping real m-tuples onto the family $\mathscr{F}_i$ of c.d.f.'s, and $\theta \equiv \{\theta_i\}_{i=1}^h$. A prior distribution $F(\theta)$ is specified, and a learning period is a sequence of times during which the designer may present the patterns of his choice to the transducer and observe the measurement z. If supervised learning may be employed, a number of questions arise. The resolution of these questions greatly influences the worth of the system. For instance, it is not generally true that the option of a learning period should be accepted. Since the object of a PRM is to recognize unknown patterns, during the supervised learning period the machine is useless in the sense that the information it is supposed to supply (which pattern is at the input) must already be available. Hence learning is feasible only if the increase in statistical knowledge counterbalances the time wasted in obtaining it. Also, the designer must decide how much of the learning period should be allotted to a particular pattern and perhaps also the total number of observations to be made. Although it is widely realized that pattern recognition is a question of optimization and also that supervised learning is an important facet of pattern recognition, the problems just mentioned have not been investigated in the engineering literature. It is the purpose of this chapter to develop a theory which provides "best" answers to these questions. It will be seen that the analytic machinery created for this purpose has effectiveness beyond these objectives: it provides an optimal method for selecting which measurements are to be made. That is, the analysis here for finding the best use of a transducer also lends itself to the problem of choosing among transducers. This latter problem is one of tremendous importance, and yet it has received scant attention in either the engineering or the statistics literature.

The first problem to be considered is the case in which the learning period is specified in advance to consist of N observations. At the k-th observation in the learning period, the designer selects from the set W some pattern which is then presented to the transducer. He records the measurement $z'_k = (z_k, w_i)$, where z_k is the transducer output at the k-th observation period and w_i is the pattern selected for that observation. The designer may consider $\{z'_i\}_{i=1}^{k-1}$ in choosing the k-th pattern. All the sets are as described in the first section of this chapter. The only additional quantity relevant to the problem is N, the length of the learning period. At the end of the learning period, the designer is to use the sample points $L = \{z'_i\}_{i=1}^{N}$ to construct a decision function which is Bayes relative to $F(\theta \mid L)$. If S is some rule for selecting patterns on the basis of past learning observations, and B(F) is the Bayes risk in recognition relative to the distribution F, then S is a solution to the supervised learning problem if

$$E[B(F(\theta \mid L)) \mid S] \leqslant E[B(F(\theta \mid L)) \mid S'] \qquad (5.12)$$

for any rule S' for selecting patterns.

The supervised learning problem just described will now be identified as a truncated Markovian adaptive control process problem. The set A of decision times is $\{1, 2, \ldots, N\}$. Assume there is a function Q and an n-tuple x_1 available to the designer such that, for $1 \leqslant j \leqslant N$,

$$Q(x_j, z'_j) = x_{j+1} \in R^n \qquad (5.13)$$

and, further,

$$F(\theta \mid \{z'_1\}_{i=1}^{j-1}) = F(\theta \mid x_j), \quad \text{and} \quad F(\theta) = F(\theta \mid x_1). \qquad (5.14)$$

Then Range (Q) will be the state-space. If nothing else, $x_{j+1} = \{z'_i\}_{i=1}^{j}$ satisfies the above requirements and therefore no PRM escapes this approach.

For each decision time, the policy set is W. The law of motion is a consequence of the definition of state-space and the PRM parameters

$$F(x_j \mid w_i, x_{j-1}) = P[C \mid w_i, x_{j-1}], \qquad (5.15)$$

where

$$C = \{z : k\text{-th component } Q(x_{j-1}, (z, w_i)) < k\text{-th component } x_j,$$
$$1 \leqslant k \leqslant V\} \quad (5.16)$$

and

$$P[C \mid w_i, x_{j-1}] = \int_C \int_{R^V} dF(z \mid \theta_i) \, dF(\theta_i \mid x_{j-1}). \qquad (5.17)$$

Because

$$F(\theta \mid x_j) = F(\theta \mid \{z'_i\}_{i=1}^{j-1})$$
$$= F(\theta \mid \{z'_1\}_{i=1}^{j-1}\{z'_1\}_{i=1}^{j-2} \cdots \{z'_1\})$$
$$= F(\theta \mid x_j, x_{j-1}, \ldots, x_2), \qquad (5.18)$$

the law of motion has the Markov property with respect to states.

Let $\bar{z}_j$ denote $\{z'_k\}_{k=1}^{j-1}$. It is computationally useful to know that, if $t(\bar{z}_j) = x_j$ is a sufficient statistic for θ' and $F(z \mid x_j, w_i, \theta)$ has a probability density function $f(z \mid x_j, w_i, \theta)$, then (5.14) holds and the law of motion has the Markov property. In view of equations (5.17) and (5.18), the assertion follows the demonstration that

$$f(\theta \mid t(\bar{z}_i)) = f(\theta \mid x_i) = f(\theta \mid \bar{z}_i).* \qquad (5.19)$$

In the analysis to follow, reference to the pattern w_i is repressed as the control element (pattern choice) is presumed fixed. By the factorability criterion for sufficient statistics [4, p. 355],

$$f(\bar{z} \mid x, \theta) = g(\theta, t(\bar{z}))h(\bar{z})$$

where $h(\bar{z})$ does not depend on θ. By Bayes rule

$$f(\theta \mid \bar{z}) = \frac{f(\theta)g(\theta, t(\bar{z}))h(\bar{z})}{\left[\int g(\theta, t(\bar{z}))f(\theta) \, d\theta\right]h(\bar{z})}$$
$$= \frac{f(\theta)g(\theta, t(\bar{z}))}{\int g(\theta, t(\bar{z}))f(\theta) \, d\theta}$$
$$\equiv \psi(\theta, t(\bar{z})). \qquad (5.20)$$

---

* For simplicity of exposition, the distributions of θ are assumed to be described by probability densities. Our method of proof is valid in the general case, however.

It follows from the abstract definition of conditional probability (for instance, see Loève [3, Section 27]) that $\psi(\theta, t(\bar{z}))$ is the conditional probability density for $\theta \mid t(\bar{z})$ if, for all events $A \times B$ in the product space $R^m \times R^n$,

$$P[A \times B] = \int_A \int_B \psi(\theta, t) \, dF_T(t) \, d\theta, \qquad (5.21)$$

F_T being the distribution induced by the function t on the random variable $\bar{Z}$. Thus

$$P[A \times B] = P\left[A \times t^{-1}(B)\right] = \int_A \int_{t^{-1}(B)} f(\theta \mid \bar{z}) \, dF_{\bar{Z}}(\bar{z}) \, d\theta$$

$$= \int_A \int_{t^{-1}(B)} \psi(\theta, t(\bar{z})) \, dF_{\bar{Z}}(\bar{z}) \, d\theta$$

$$= \int_A \int_B \psi(\theta, t) \, dF_T(t) \, d\theta.$$

The last equation involves a transformation of variables in integration, and the preceding equation (5.21) includes a substitution allowed by (5.20). Therefore $f(\theta \mid t(\bar{z})) = f(\theta \mid \bar{z})$ and (5.19) is correct.

If all sets $\mathscr{F}_i$, $1 \leqslant i \leqslant h$, are composed entirely of either discrete or absolutely continuous c.d.f.'s it is now shown that $B(F(\theta \mid x_{t+1}))$ is a well-defined quantity that may be found from the process sets. Following the definition in Blackwell and Girshick, the rule d^* is Bayes with respect to $F(\theta \mid x_{t+1})$ if, for each z ([1, Chapter 6]),

$$d^*(z) = w^* \in W \qquad (5.22)$$

such that

$$K(z, w^*) = \min_{w \in W} K(z, w), \qquad (5.23)$$

where

$$K(z, w) = \sum_{w' \in W} H(w', w) P(w' \mid z, x_{t+1}). \qquad (5.24)$$

In the case that the functions in $\mathscr{F}_i$ are discrete, by Bayes rule,

$$P(w_i \mid z, x_{t+1}) = \frac{P(w_i) P(z \mid w_i, x_{t+1})}{\sum\limits_{w' \in W} P(w') P(z \mid w', x_{t+1})} \qquad (5.25)$$

where

$$P(z \mid w_j, x_{t+1}) = \int_{R^m} P(z \mid \theta_j) \, dF(\theta_j \mid x_{t+1}). \qquad (5.26)$$

By definition,

$$B\big(F(\theta \mid x_{t+1})\big) = \sum_{z, P(z) > 0} K\big(z, d^*(z)\big)P(z) \qquad (5.27)$$

$$P(z) = \sum_{w' \in W} P(z \mid w', x_{t+1})P(w'). \qquad (5.28)$$

In the case that all members of $\mathscr{F}_i$, $1 \leqslant i \leqslant h$, are absolutely continuous, the analysis is different. If B_i is the closed interval

$$B_i = [z - (1/i), z], \qquad (5.29)$$

then by the constructive definition for conditional distributions [4, p. 60],

$$
\begin{aligned}
P(w \mid z, x_{t+1}) &= \lim_{i \to \infty} \frac{P(w, B_i, x_{t+1})}{P(B_i, x_{t+1})} = \lim_{i \to \infty} P(w \mid B_i, x_{t+1}) \\
&= \lim_{i \to \infty} \frac{P(w)P(B_i \mid x_{t+1}, w)}{\sum_{w' \in W} P(w')P(B_i \mid x_{t+1}, w')} \\
&= \lim_{i \to \infty} \frac{P(w)\big[F(z \mid x_{t+1}, w) - F\big(z - (1/i) \mid x_{t+1}, w\big)\big]}{\sum_{w' \in W} P(w')\big[F(z \mid x_{t+1}, w') - F\big(z - (1/i) \mid x_{t+1}, w'\big)\big]} \\
&= \lim_{i \to \infty} \frac{P(w)\left[\dfrac{F(z \mid x_{t+1}, w) - F\big(z - (1/i) \mid x_{t+1}, w\big)}{(1/i)}\right]}{\sum_{w' \in W} P(w')\left[\dfrac{F(z \mid x_{t+1}, w') - F\big(z - (1/i) \mid x_{t+1}, w\big)}{(1/i)}\right]} \\
&= \frac{P(w)f(z \mid x_{t+1}, w)}{\sum_{w' \in W} P(w')f(z \mid x_{t+1}, w')}, \qquad (5.30)
\end{aligned}
$$

where

$$f(z \mid x_{t+1}, w_i) = d/dz \int F(z \mid \theta_i) \, dF(\theta_i \mid x_{t+1}).^* \qquad (5.31)$$

It follows from the Radon-Nikodyme theorem [3, Section 8] that $B\big(F(\theta \mid x)\big)$ is a well-defined quantity for PRM's having arbitrary families $\mathscr{F}_i$ of c.d.f.'s. Unfortunately, under general circumstances there is no effective procedure for determining $B\big(F(\theta \mid x)\big)$.

If

$$
L(z, y, w, j) = \begin{cases} 0, & \text{if } j < N, \\ B\big(F(\theta \mid y)\big), & \text{if } j = N, \end{cases} \qquad (5.32)
$$

then, referring to the theory of Chapter 3 it is clear that the loss function is

---

* This analysis extends without difficulty to n-dimension c.d.f.'s.

separable, and with this remark the identification of this supervised learning problem as a truncated Markovian adaptive control process problem has been completed. One may further conclude that, if a solution S exists, it may be found by dynamic programming and must have the property that

$$E[B(F(\theta/x)) \mid S] \leqslant E[B(F(\theta/x)) \mid S'] \qquad (5.33)$$

for all other strategies S', as required.

In summary, a PRM with an N-stage learning period is characterized by the pattern set $W = \{w_1, w_2 \ldots, w_h\}$, the functions $\boldsymbol{\theta}_i$ and $F(\theta)$, the sets $\mathscr{F}_i$ of c.d.f.'s, and loss matrix H. In modeling the problem of using the learning period to minimize the Bayes risk at the termination of learning as a truncated Markovian ACP problem, the set of decision times A is the positive integers through N, the control set at all state-times is W, and the state-space is the range of a function Q with the properties of (5.13) and (5.14). The law of motion is determined from the sets $\mathscr{F}_i$ and the function $F(\theta)$ as indicated in equations (5.15), (5.16), and (5.17). Finally, the loss function is given in terms of the Bayes risk $B(F(\theta \mid x))$ by the relation(5.32). $B(F(\theta \mid x))$ is uniquely determined by the PRM sets, and, in the event that the sets $\mathscr{F}_i$ are composed either of absolutely continuous or discrete c.d.f.'s, there is an effective way to compute the Bayes risk.

If digital computation is used, all sets must be finite. Under this condition, Theorem 3.3 assures that a solution exists and dynamic programming will find it efficiently.

It is interesting to note that the supervised learning problem just investigated has the same mathematical structure as the bandit problem B_1 of Chapter 4, in which the player is allowed to keep only the payoff at the final play.

Using the identification of pattern recognition quantities as adaptive control process sets made in the preceding analysis, it is possible to deal with all the questions concerning supervised learning which were mentioned at the beginning of this section. For example, suppose that the number of learning observations is left to the designer's discretion, but that the total life of the transducer is N observations. A strategy is required that not only selects patterns for learning observations as above, but also specifies, on the basis of past observations, when the learning period should be terminated and the machine put to use. This strategy must give no higher expected loss than any other strategy. In short, there must be no way of using the machine which yields a greater expected worth than the solution called for. It is conceivable that the solution will call for never using the machine (this will happen, for instance, if H is positive for all arguments). At the other extreme, it is possible that the solution will require that there be no learning observations.

The identification of state-space and the law of motion for this problem is the same as before. However, the policy set at each period includes an additional element p', the stopping option. If the designer elects to begin recognition operation (policy p') at the k-th decision time, then the total loss during the rest of the machine's life is $(t - k + 1) B(F(\theta/x_k))$. That is, at each usage, the risk is $B(F(\theta/x_k))$, and $(N - k + 1)$ usages remain. If learning is continued, no loss is incurred. From this description, one concludes that the loss function is specified by

$$L(x, y, p, j) = \begin{cases} (N - j + 1)B(F(\theta \mid x)), & \text{if } p = p', \\ 0, & \text{if } p \in W. \end{cases} \quad (5.34)$$

A simple numerical problem of the sort just discussed serves to illustrate the ideas of this chapter.

EXAMPLE. $N = 6$; $W = \{A, B\}$; $p(w) = \frac{1}{2}$, $w = A, B$.
For Pattern A,

$$p(z \mid \theta_A, A) = \begin{cases} \frac{1}{4} \text{ for } z = 0, \frac{3}{4} \text{ for } z = 1, & \text{if } \theta_A = 1, \\ \frac{3}{4} \text{ for } z = 0, \frac{1}{4} \text{ for } z = 1, & \text{if } \theta_A = 2. \end{cases}$$

For Pattern B,

$$p(z \mid \theta_B, B) = \begin{cases} \frac{1}{4} \text{ for } z = 0, \frac{3}{4} \text{ for } z = 1, & \text{if } \theta_B = 1, \\ \frac{3}{4} \text{ for } z = 0, \frac{1}{4} \text{ for } z = 1, & \text{if } \theta_B = 2. \end{cases}$$

$p[\theta_A = 1] = p[\theta_A = 2] = \frac{1}{2} = p[\theta_B = 1] = p[\theta_B = 2]$.

The recognition loss matrix is

		Actual pattern	
H(w, g):	g/w	A	B
Guess	A	-1	1.5
	B	1.5	-1

(5.35)

where w_i denotes the pattern actually at the transducer, and g_j the guess. Figure 5.2 displays a solution to this example. The computational equations and computer program and printout for this problem comprise Appendix I.

The mathematical structure of this last supervised learning problem is identical to the bandit with the quitting option, bandit B_2, of Chapter 4. The fundamental game may be viewed as one wherein the loss function is

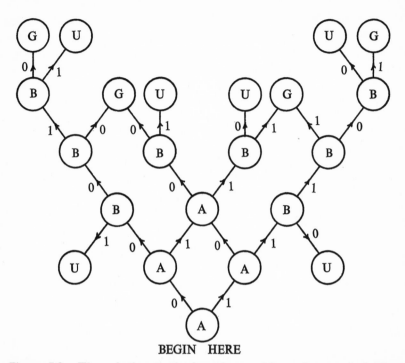

BEGIN   HERE

Figure 5.2.   The solution to the learning problem of example I. The node indicates the action to be taken and the branch with the label of the resulting outcome leads to the next decision. A means "Learn A," B "Learn B," U "Use," and G "Give Up."

0 for all arguments. Superimposed on this game is the house offer of $((t - j + 1)B(F(\theta \mid x))$ at each stage j, the system being in state x, if the player agrees to quit the game.

3. GENERALIZATIONS OF THE SUPERVISED LEARNING PROBLEM

It is interesting to consider the sort of supervised learning problems that can be solved by dynamic programming. Given the identification of state and policy spaces and the law of motion, as considered in the analysis of the last section, any problem yielding a loss function of the form

$$L(\bar{x}, \bar{p}, t) = \sum_{i=t}^{N} L(\bar{x}(i), \bar{x}(i+1), \bar{p}(i), i)$$

is permissible. It is immediate that a cost $C(w_i, k)$ for choosing to test pattern w_i for the k-th observation may be levied without affecting the problem's Markovian nature.

The loss function of the past analysis,

$$(N - j + 1)B(F(\theta \mid x)) = L(x, y, p, j),$$

is a functional on $F(\theta \mid x)$. Any other functional on $F(\theta \mid x)$ will also yield a truncated Markovian adaptive control process. As transinformation is also a functional on $F(\theta \mid x)$, it is possible to devise a learning scheme to maximize the transinformation of a channel. For any particular decision function d, the transinformation may be calculated [6, p. 105], given the PRM sets and the final state $x(N + 1)$. Let w_i denote the actual pattern and $v_j \in W$, the guess, then

$$I_d(W, V) = \sum_{i=1}^{h} \sum_{j=1}^{h} P(w_i, v_j) \log\left(P(w_i, v_j)/P(w_i)P(v_j)\right). \qquad (5.36)$$

For a particular d,

$$P(w_i, v_j) = \int_{Q_j}\int_{R^M} dF(z \mid \theta_i)\, dF(\theta_i \mid x(N + 1))P(w_i), \qquad (5.37)$$

where

$$Q_j = d^{-1}(v_j). \qquad (5.38)$$

$$P(v_j) = P[z \in Q_j] = \sum_{w' \in W} P(w', v_j). \qquad (5.39)$$

Hence $I_d(W, V)$ is readily calculated from the known quantities. It is natural to define the channel transinformation to be

$$I(W, V) = \sup_{d} I_d(W, V) = \psi(F(\theta \mid x(N + 1))). \qquad (5.40)$$

Then by defining the gain function to be

$$G(x, y, p, j) = \begin{cases} 0, & j < N, \\ \psi(F(\theta \mid y)), & j = N, \end{cases} \qquad (5.41)$$

it is evident that a solution to this problem yields the greatest possible expected transinformation. If the designer has the capability of specifying the pattern input probabilities $P(w_i)$, then, in the manner of maximizing the channel transinformation, he may use the learning period to maximize the resulting channel capacity, as defined by Shannon [6, p. 109]. The problem of maximizing the channel capacity is solved by the preceding analysis with a slight change in loss function. Now we have

$$G(x, y, p, j) = \begin{cases} 0, & j < N, \\ \max_{\xi \in \Xi} \Psi'_\xi(F(\theta \mid y)), & j = N, \end{cases}$$

where $\xi = (\xi_1, \ldots, \xi_h)$ and $\Psi'_\xi(F(\theta \mid y))$ is the channel transinformation as defined in (5.40) if $p(w_1) = \xi_1, \; p(w_2) = \xi_2, \ldots, p(w_h) = \xi_h$. Ξ is the convex hull of the set of the set of h-tuples

$$\{(1, 0, \ldots, 0), (0, 1, \ldots, 0), \ldots, (0, 0, \ldots, 1)\}.$$

Historically, the researchers studying PRM's have maintained a strictly Bayesian statistical viewpoint. However, the theory of this chapter is applicable no matter what the method of estimation. Since the concept of risk is customarily associated only with Bayes estimation, some other criterion of performance for PRM's would have to be accepted.

4. OPTIMUM TESTING FOR THE SELECTION OF A TRANSDUCER

In this section it is supposed that several different transducers are available. During the learning period, the designer can elect to have *any* pattern presented to *any* transducer. At the termination of learning, the designer selects the transducer yielding the minimum Bayes risk for use in the PRM. He wishes to find some method for testing the transducers so that he may expect that, at the termination of learning, the PRM will perform as well as possible.

An example in which this problem might occur is in the design of a machine to recognize printed letters by means of an outcome z resulting from a machine measurement on a grid of photoelectric cells. Perhaps there are two possible ways of obtaining z. If the first method is used, z is some function of the degree of curvature in the lines of the measured letter. Alternatively, the transducer may be adjusted so that z is dependent on the relative intensity at various zones on the grid. Due to limitations on the complexity of the transducer, both measurements cannot be made at once. Hence, at the end of the learning period the designer must choose one way or the other for PRM operation.

To formalize this problem, a transducer is defined by the triple

$$T = \left(\{\mathscr{F}_i\}_{i=1}^h, F(\theta), \boldsymbol{\theta}\right), \tag{5.42}$$

where the elements of T are as defined in Section 2. It is assumed that the designer has α different transducers available to him, and he is allowed a learning period of N observations to be distributed among the machines and patterns as he wishes. A strategy S based on past observations is a solution to the transducer testing problem if, letting $B_j(F)$ denote the Bayes risk with respect to F of the j-th transducer and L the sequence of learning observations,

$$E\left[\min_{1\leqslant j\leqslant \alpha} B_j(F(\theta/L)) \mid S\right] \leqslant E\left[\min_{1\leqslant j\leqslant \alpha} B_j(F(\theta/L)) \mid S'\right] \tag{5.43}$$

for any other strategy S' on the past learning observations.

In the identification of the transducer selection problem as an adaptive control process problem, for all states and decision time, the control set will be taken to be

$$P = \{(j, w_i) : 1 \leqslant j \leqslant \alpha, 1 \leqslant i \leqslant h\}, \tag{5.44}$$

where (j, w_i) indicates pattern w_i is to be presented to transducer j.

Assume that each of the transducers has been identified as an adaptive control process according to the discussion earlier in this chapter, and Ω_j is the state-space of the j-th transducer. Then the state-space of the transducer problem will be:

$$\Omega' = \{x' = (x_1, x_2, \ldots, x_\alpha) : x_j \in \Omega_j, 1 \leqslant j \leqslant \alpha\}. \tag{5.45}$$

The state x' indicates that, for each j, $1 \leqslant j \leqslant \alpha$, transducer j is in state x_j.

The law of motion follows logically from the above definitions and the assumption that the transducers are statistically independent of one another. Let $y' = (y_1, y_2, \ldots, y_\alpha)$ and $x' = (x_1, x_2, \ldots, x_\alpha)$. Since y_k depends on the statistical law of the k-th transducer and not the others, by the assumption of transducer independence,

$$F(y' \mid x', (j, w)) = \prod_{k=1}^{\alpha} F(y_k \mid x', (j, w)). \tag{5.46}$$

Again from the independence assumption,

$$F(y_k \mid x', (j, w)) = F(y_k \mid x_k, (j, w)). \tag{5.47}$$

If $j \neq k$, then $F(y_k \mid x_k, (j, w))$ is the n-dimensional unit step function at x_k, because the k-th transducer is not used and therefore its state is unchanged. If $j = k$,

$$F(y_k \mid x_k, (k, w)) = F(y_k \mid x_k, w),$$

which is an element of the law of motion of the k-th machine and therefore a known quantity. Hence the statistical law of motion is strictly determined for the transducer selection problem. That this law has the Markov property with respect to states follows from the fact that the transducer laws individually have the Markov property.

Given any state $x' = (x_1, x_2, \ldots, x_\alpha)$, the state of the k-th transducer is x_k, and if the k-th transducer is chosen for recognition operation the Bayes risk will be $B(F(\theta \mid x_k))$, as discussed in preceding sections. Letting $B_k(F)$ denote the Bayes risk if the k-th transducer is used,

$$B_k(F(\theta \mid x')) = B_k(F(\theta \mid x_k))$$

is defined, given x', for $1 \leqslant k \leqslant \alpha$. Therefore

$$B'(x') = \min_{1 \leqslant k \leqslant \alpha} \{B_k(F(\theta \mid x'))\} \qquad (5.48)$$

is well defined, and represents the Bayes risk of the PRM using the transducer which *a posteriori* performs best. The following separable loss function completes the truncated Markovian adaptive control process description:

$$L(x', y', p, k) = \begin{cases} B'(y'), & \text{if } k = N, \\ 0, & \text{if } k \neq N. \end{cases} \qquad (5.49)$$

By the theory in Chapter 3, dynamic programming may be used to obtain a solution S which satisfies the relation

$$E[B'(y') \mid S] \leqslant E[B'(y') \mid S'], \qquad (5.50)$$

y' being the $N + 1$-*st* state, for all other strategies S'. Except for change in nomenclature, the above is the same as the relation (5.43) defining a solution to the transducer selection problem.

One may wonder whether there is generally any advantage to having a number of transducers to test over just having one transducer, and testing it thoroughly. Does the increase in complication generally result in better performance? The answer is affirmative, because the strategy of testing only one machine is among the set of alternatives, and if it yields the least expected risk, by (5.50), it will be selected.

As with the learning problem, the transducer selection problem admits very general extensions. In essentially the same way as in the preceding section, for instance, if the total number of observations (learning plus recognition) is fixed, the loss function may be modified to give a stopping rule as well as the learning selections.

If the state-space for the transducer selecting problem is as discussed above, but in addition displays the number of the last transducer tested, then the loss function may be modified to attach a cost for changing transducers during the testing period.

The methods of this chapter are applicable to the problem of selecting the best machine in a set of different PRM's. In this situation, at each stage of a learning period, the designer presents a pattern to a machine (both chosen at his discretion) and observes the guess.

5. HISTORY AND SUMMARY OF PATTERN RECOGNITION THEORY

In the late 1940's Claude Shannon [7] and Norbert Wiener [8], from different points of view, provided the nucleus for statistical communication theory. Roughly, classic statistical communication theory is concerned with how a received signal should be treated to recover as well as possible the message sent. It is assumed that this message has been distorted by the addition of a noise signal whose statistical characteristics are known.

In Wiener's framework, the message and noise functions are defined on a continuum of points and have certain very restrictive characteristics. Moreover, the analysis of the received signal must be done by linear networks (filters).

Shannon's approach, more akin to the pattern recognition machine formulation, assumes that at each time in a countable set of decision times a member is chosen from a finite set of patterns as input to a transducer (communication channel). As in Wiener's framework, Shannon assumes that the statistical characteristics of the channel are known. The designer seeks to specify the way patterns should be selected to maximize a particular gain function called "information rate."

Learning problems arise in generalizing Shannon's model to allow that channel characteristics (i.e., transducer statistics) are not known in advance. (Also, more general loss functions are permitted.) Abrahamson and Braverman [9] give a good bibliography on the early literature of learning. An up-to-date comprehensive bibliography is to be found in the paper by Sklansky [10].

Braverman [11], Spragins [12], and other researchers under Abrahamson's direction, exploiting Bayesian statistical decision theory, analyzed how best to use data obtained during a learning period, assuming the learning observations had already been made, to minimize Bayes risk in recognition when the machine is put into operation.

As a result of practical considerations, most studies in the design of a supervised learning period have been specialized to apply only to a restricted class of machines (linear adaptive threshold devices) for which decision boundaries on the sample space of transducer outputs are hyperplanes. The analysis here is generally non-parametric, and researchers dealing with these machines are seldom concerned with optimal use of the learning period, being content to show that the learning algorithms work at all. Nilsson's book [13] affords a comprehensive introduction to the subject of linear adaptive threshold devices. This author believes that modern digital computers can feasibly be used as pattern recognition machines with more general decision regions wherein sequential decision theory can profitably be employed.

Two papers on supervised learning in a general context are known to this author. Highleyman [14] considers how a learning period should be divided between learning and evaluation phases. The method he proposes for designing the learning period depends not on observations during the learning period but entirely on the prior distributions and thus no sequential analysis is called for. Chen [15] proposes a sequential learning technique which has intuitive appeal, but he does not supply logical justification.

We mention some other studies that use DP in analyzing problems of statistical communication theory. In "Functional Equations in Adaptive Processes and Random Transmission," Bellman and Kalaba [16] show how DP can be used to find the electromagnetic transmission properties of an inhomogeneous medium having a randomly varying index of refraction. Their paper "On Communication Processes Involving Learning and Random Duration" [17] contains a general discussion of control and communication under the circumstance that the statistical knowledge of the process is not complete. It gives some space to a problem first studied by Kelley (see the reference of Problem 7, of Section 2, Chapter 3) and referred to as "the gambler with the noisy wire." In this problem the communication-theoretical concept of information occurs in an economic setting. In [17], the problem has been generalized so that learning and control must be performed simultaneously by "the adaptive gambler." Also, there is a brief analysis of the case in which the length of the process is a r.v.

The paper entitled "Dynamic Programming, Sequential Estimation and Sequential Detection Processes" [18] gives a summary of a slightly restricted version of Wald's sequential decision problem. The restriction is such that DP is now applicable, and this paper shows how the application is to be made. The paper further indicates the relevance of classic sequential decision theory to problems of communication theory, and mention is made of the computational difficulties encountered in viewing communication problems as sequential decision problems.

The book *Dynamic Programming and Modern Control Theory* [19] suggests (Chapter IV) several applications of DP to major engineering questions in control and communication theory. The book is a fine source of references for papers in this area and points out avenues for future research.

In addition to the material based on the Glivenko-Cantelli lemma on the rate of convergence in learning, the contribution of this chapter has been to give a general procedure for constructing a strategy for a learning period so that the expected Bayes risk in recognition at the termination of learning will be minimized. We have seen that the supervised learning problem is also an adaptive control process problem. Among the generalizations resulting from this insight is that dynamic programming is useful in choosing a transducer, or equivalently selecting among possible ways of performing pattern recognition.

Exercises on Learning in Pattern Recognition Theory

(1) Let the PRM parameters W, $\{(\mathscr{F}_i, \theta_i)\}_{i=1}^{h}$, $F(\theta)$, and H, as defined in Sections 1 and 2, be specified. The situation in this problem differs from the learning processes described in Section 2 in that, at the termination of learning, the sequence $\{p(w_i)\}_{i=1}^{h}$ which displays the relative frequency that the patterns are presented to the transducer is to be chosen at the discretion of the designer from a set B of sequences. Under these circumstances, indicate how to find a strategy for using an N-stage learning period so that the recognition risk per observation at the termination of learning will be minimized. Show that, if B is the set of all sequences of probabilities which sum to 1, then, provided the loss for misclassification is higher than for recognition, a Bayes solution for using the PRM will call for assigning probability 1 to some pattern.

(2) Consider the problem of the gambler with the noisy communication channel, Problem 7, Section 2, Chapter 3. Suppose he does not know the value of p (which is a constant throughout the betting process) in advance, but knows that it has been selected by a random experiment whose c.d.f. is G(p). Determine, under these circumstances, how the gambler should bet so as to maximize the expected log of his fortune at the end of his N-th wager. See reference [1, p. 143] of Chapter 3.

(3) In Section 3 the question is discussed of using a learning period to maximize channel capacity. Suppose that the PRM quantities W, $\{\mathscr{F}_i\}_{i=1}^{h}$, θ, $F(\theta)$, and N, the learning period length, are specified as in the set-up of Section 3. The difference is that now a cost c is levied for each letter w_i such that $P(w_i) \neq 0$. This cost is subtracted from the channel capacity to determine the value of the machine at the termination of learning. Indicate how to find a learning strategy that maximizes this value.

(4) If a PRM has been designed and programmed, it may be described by the h × h matrix $P = [p_{ij}]$, where p_{ij} is the probability that if pattern w_i is presented to the transducer the machine will "guess" w_j. Assume that the PRM quantities W, H, and $\{p(w_i)\}_{i=1}^{h}$ are given, but the matrix P is not known. However, in Bayes fashion, *a priori*, distribution F(P) for the elements

in P is available. The designer is to use an N-stage learning period for accept-ance testing. At each decision period, on the basis of previous observations, he can select a pattern to be presented to the PRM, and observe the resulting guess. The machine is to be accepted if and only if the expected loss per observation is less than β. If the designer makes the wrong acceptance decision, he experiences a loss of 1, whereas his loss is 0 if he is correct. How should he select the patterns during the learning period in order to minimize his expected loss when he makes his acceptance decision, and how is the acceptance decision to be determined from the learning observations?

REFERENCES

1. Blackwell, D., and M. A. Girschick (1954). *Theory of Games and Statistical Decisions.* Wiley, New York.
2. Aoki, M. (1965). On some convergence questions in Bayesian optimization problems. *IEEE Trans. Automatic Control 10*, 180–182.
3. Loève, M. (1963). *Probability Theory*, 3rd ed. Van Nostrand, Princeton, N.J.
4. Wilks, S. S. (1962). *Mathematical Statistics.* Wiley, New York.
5. Apostol, T. M. (1957). *Mathematical Analysis.* Addison-Wesley, Reading, Mass.
6. Reza, F. M. (1961). *An Introduction to Information Theory.* McGraw-Hill, New York.
7. Shannon, C. (1948). A Mathematical Theory of Communication. *Bell System Tech. J. 27*, 379–423, 623–656.
8. Wiener, N. (1949). *The Extrapolation, Interpolation and Smoothing of Stationary Time Series with Engineering Applications.* Wiley, New York.
9. Abrahamson, N., and D. Braverman (1962). Learning to recognize patterns in a random environment. *IEEE Trans. Information Theory 8*, 58–63.
10. Sklansky, J. (1966). Learning systems for automatic control, *IEEE Trans. Automatic Control 11*, 6–19.
11. Braverman, D. (1961). *Machine Learning and Automatic Pattern Recognition.* T.R. 2003-1, Stanford Electronic Laboratories, Stanford, Calif.
12. Spragins, J. (1963). *Reproducing Distributions for Machine Learning.* T.R. 6103-7, Stanford Electronic Laboratories, Stanford, Calif.
13. Nilsson, N. (1965). *Learning Machines.* McGraw-Hill, New York.
14. Highleyman, W. (1962). The design and analysis of pattern recognition experiments. *Bell System Tech. J. 41*, 723–744.
15. Chen, C. (1966). A note on the sequential decision approach to pattern recognition and learning. *Information and Control 9*, 549–562.
16. Bellman, R., and R. Kalaba (1959). Functional equations in adaptive processes and random transmission. *Trans. 1959 Intern. Symp. on Circuit & Information Theory*, 271–282.
17. Bellman, R., and R. Kalaba (1958). On communication processes involving learning and random duration. *1958 IRE National Convention Record.* Part 4, 16–20.
18. Bellman, R., R. Kalaba, and D. Middleton (1961). Dynamic programming, sequential estimation and sequential detection processes. *Proc. Nat. Acad. Sci. U.S.A. 47*, 338–341.
19. Bellman, R., and R. Kalaba (1966). *Dynamic Programming and Modern Control Theory.* Academic Press, New York.

THEOREM ON THE CONTINUITY
OF AN INTEGRAL

LEMMA.   *Let* $g(x_1, x_2)$ *be a continuous function defined on* $R^n \times R^m$, *and* C *a compact subset of* R^m. *Then the function*

$$u(x_1) = \int_C g(x_1, x_2)\, dx_2$$

is continuous.

PROOF.   Let $\{x_1(n)\}$ be any sequence converging to a point x_1 in R^n, and S a closed n-dimensional sphere containing this sequence. Then $S \times C$ is a compact set and $|g|$ is therefore bounded by a real number M, on $S \times C$ [1]. From the continuity of g, we have

$$\lim_{n \to \infty} g(x_1(n), x_2) = g(x_1, x_2)$$

for every point x_2. Thus Lebesgue's bounded convergence [2] theorem is applicable:

$$\lim_{n \to \infty} u(x_1(n)) = \int_C \lim_{n \to \infty} g(x_1(n), x_2)\, dx_2$$

$$= \int_C g(x_1, x_2)\, dx_2 = u(x_1).$$

REFERENCES

1. Apostol, T. M. (1957). *Mathematical Analysis*, p. 73. Addison-Wesley, Reading, Mass.
2. Halmos, P. R. (1950). *Measure Theory*, p. 110. Van Nostrand, Princeton, N.J.

THEOREM ON THE STRUCTURE OF SOLUTIONS FOR TABS WITH ONE MACHINE KNOWN

LEMMA.   *If* S *is a solution to a TAB problem of Section 1, Chapter 4, and for some state-time pair* (x, t) *occurring with positive probability,* S(x, t) = II, *then* E[G(x, S, t)] = (N − t + 1)s.

PROOF.   The proof is by induction on $b = N - t + 1$, the number of decision times remaining at time t. For $b = 1$, the assertion is immediate since

$$E[G(x, S, N)] = E[g(x, Y) \mid x, II] = s.$$

Suppose the assertion is true if $b < \xi$. Let $t = N - \xi + 1$, and suppose S(x, t) = II for some solution S. For expository reasons, it is helpful to consider two problems arising from the TAB process:

Game A, The initial state is x and there are ξ decision times.

Game B, The initial state is x and there are $\xi - 1$ decision times.

With this terminology, it is evident that the modified problem at (x, t) is Game A and the problem at (x, t + 1) is Game B. In view of Lemma 4.1, if x′ differs from x only in the third component, then the modified problem at (x′, t + 1) is also Game B. Thus as S(x, t) = II, Game B must occur at time t + 1 under strategy S if (x, t) occurs. By the principle of optimality theorem, Theorem 3.4, S, being a solution, must also be a solution to the modified problems occurring with positive probability, and in particular to the modified problems at (x, t) and (x, t + 1). Thus

$$value(Game\ A) = E[G(x, S, t)]$$

$$= E\big(g(x, Y) \mid x, S, t\big) + E[G(Y, S, t + 1) \mid x, S, t] \quad (1)$$

$$= s + value(Game\ B).$$

$\overline{X}$ will denote the trajectory beyond (x, t) under strategy S. The next objective in this proof is to demonstrate that

$$E[g(\overline{X}(N), \overline{X}(N + 1) \mid x, S, t)] = s. \quad (2)$$

If (2) is not true, then for some state $y = (q_1, q_2)^*$,

$$P[X(N) = y \mid x, S, t] = c_1 > 0 \quad \text{and} \quad r(q_1, q_2) - s = c_2 > 0,$$

and consequently S(y, N) = I.

_____

* In view of Lemma 4.1, our attention is restricted to the first two state components.

It follows that

$$P[X(N + 1) = (q_1 + 1, q_2) \mid x, S, t]$$
$$= P[X(N) = (q_1, q_2) \mid x, S, t]P[(q_1 + 1, q_2) \mid (q_1, q_2), I] \qquad (3)$$
$$= c_1 r(q_1, q_2).$$

In [1, p. 226] it is seen that

$$r(q_1 + 1, q_2) > r(q_1, q_2), \qquad (4)$$

and we shall need this fact shortly. Let the strategy S' be defined so that

$$S'(z, j) = \begin{cases} S(z, j + 1), & t \leqslant j < N, \\ II, & j = N \text{ and } z \neq (q_1 + 1, q_2), \\ I, & j = N \text{ and } z = (q_1 + 1, q_2). \end{cases}$$

The first $\xi - 1$ plays beyond state-time (x, t) is Game B, as is the modified problem at $(x, t + 1)$. We have mentioned that S is a solution to the modified problem at $(x, t + 1)$. S', at times less than N, is S translated one time unit earlier. Thus

$$E\left[\sum_{i=t}^{N-1} g(\overline{X}(i), \overline{X}(i + 1)) \mid x, S', t\right] = E\left[\sum_{i=t+1}^{N} g(\overline{X}(i), \overline{X}(i + 1)) \mid x, S, t + 1\right]$$
$$= \text{value (Game B)}.$$

Also,

$$P[X(N + 1) = (q_1 + 1, q_2) \mid x, S, t]$$
$$= P[X(N) = (q_1 + 1, q_2) \mid x, S,' t]$$
$$= r(q_1, q_2)c_1 > 0.$$

Therefore,

$$E[g(X(N), X(N + 1)) \mid x, S', t]$$
$$= s(1 - r(q_1, q_2)c_1) + r(q_1, q_2)c_1(r(q_1 + 1, q_2)).$$

Recalling (4), this implies $E[g(X(N), X(N + 1)) \mid x, S', t] > s$. In summary,

$$E[G(x, S', t)] > \text{value(Game B)} + s = E[G(x, S, t)],$$

contradicting the assumption that S is a solution at (x, t). In view of (1) and

(2), we have now demonstrated that

$$E\left[\sum_{i=t}^{N-1} g(X(i), X(i+1)) \,\middle|\, x, S, t\right] = \text{value (Game B)}.$$

Thus S, restricted to times $\{t, t+1, \ldots, N-1\}$ is a solution to Game B. As $S(x, t) = II$, the inductive hypothesis implies (as the number of elements of $\{t, \ldots, N-1\} = \xi - 1$) that

$$\text{value(Game B)} = (\xi - 1)s.$$

Thus from (1), we conclude value(Game A) $= \xi s$, completing the inductive proof.

REFERENCE

1. Bellman, R. (1956). A problem in the sequential design of experiments. *Sankhyā* **16**, 221–229.

Appendix C

COMPUTER PROGRAM AND READOUT
FOR TAB PROBLEM,
ONE MACHINE KNOWN

```
*     COMPILE FORTRAN,EXECUTE FORTRAN
*** MAIN PROGRAM ***
    I DEFINED BUT NOT USED IN AN ARITH STMNT.
   II DEFINED BUT NOT USED IN AN ARITH STMNT.
    N DEFINED BUT NOT USED IN AN ARITH STMNT.
             DIMENSION EGAIN(2000,2)
2002     1 FORMAT(I2,2(A2),4XF10.4)
2007     2 FORMAT(1H132H*** TWO-ARMED BANDIT PROBLEM ***/5X28HBANDIT NUMBER T
        1WO KNOWN, S= F6.4/5X24HTOTAL NUMBER OF PLAYS = I2)
2031     3 FORMAT(//4H K= I2//5X5HSTATE9X1HR8X4HH(I)7X5HH(II)6X5HEGAIN4X
        16HS(X,1)/)
2046     4 FORMAT(4X1H(I2,1H,I2,1H)4(5XF6.4),4X2(1XA2))


2053  1000 READ 1,110,NTIME,I,II,S
2063       PRINT 2,S,NTIME
2069       NC=1

'DO 100' CONTROLS THE TIME PERIOD.
2071       DO 100 IN=1,NTIME
2075       NS=0
2077       K=NTIME-IN+1
2081       PRINT 3,K
2086       KK=K-1

COMPUTE H2.
2089       CH2=NTIME-KK
2093       H2=CH2*S

'DO 70' GENERATES THE STATE SET ELEMENTS (M,N).
2096       DO 70 M=0,KK
2100       NS=NS+1
2103       N=KK-M

COMPUTE 'R',THE ESTIMATOR FOR MACHINE I.
2106       RN=M+1
2110       RD=K+1
2114       R=RN/RD

          IN FINAL DECISION PERIOD H1 IS COMPUTED AS PER STMNT. 10,

          OTHERWISE AS PER STMNT. 20.
2118           IF(IN-1)20,10,20
2122        10 H1=R
2124           GO TO 30
2125        20 H1=R*(1.+EGAIN(NS+1,LNC))+(1.-R)*EGAIN(NS,LNC)

          DETERMINE DECISION S(X,T).
2142        30 IF(H1-H2)40,50,60
2147        40 EGAIN(NS,NC)=H2
2153           PRINT 4,M,N,R,H1,H2,EGAIN(NS,NC),II
2166           GO TO 70
2167        50 EGAIN(NS,NC)=H1
2173           PRINT 4,M,N,R,H1,H2,EGAIN(NS,NC),I,II
2187           GO TO 70
2188        60 EGAIN(NS,NC)=H1
2194           PRINT 4,M,N,R,H1,H2,EGAIN(NS,NC),I
2207        70 CONTINUE
2208           GO TO(80,90),NC
2213        80 NC=2
2215           LNC=1
2217           GO TO 100
2218        90 NC=1
2220           LNC=2
2222       100 CONTINUE
2223           GO TO 1000
2224       110 STOP
               END


          MAIN PROGRAM
```

```
*** TWO-ARMED BANDIT PROBLEM ***
    BANDIT NUMBER TWO KNOWN, S= 0.6000
    TOTAL NUMBER OF PLAYS =  6
```

K= 6

STATE	R	H(I)	H(II)	EGAIN	S(X,1)
(0, 5)	0.1429	0.1429	0.6000	0.6000	II
(1, 4)	0.2857	0.2857	0.6000	0.6000	II
(2, 3)	0.4286	0.4286	0.6000	0.6000	II
(3, 2)	0.5714	0.5714	0.6000	0.6000	II
(4, 1)	0.7143	0.7143	0.6000	0.7143	I
(5, 0)	0.8571	0.8571	0.6000	0.8571	I

K= 5

STATE	R	H(I)	H(II)	EGAIN	S(X,1)
(0, 4)	0.1667	0.7667	1.2000	1.2000	II
(1, 3)	0.3333	0.9333	1.2000	1.2000	II
(2, 2)	0.5000	1.1000	1.2000	1.2000	II
(3, 1)	0.6667	1.3429	1.2000	1.3429	I
(4, 0)	0.8333	1.6667	1.2000	1.6667	I

K= 4

STATE	R	H(I)	H(II)	EGAIN	S(X,1)
(0, 3)	0.2000	1.4000	1.8000	1.8000	II
(1, 2)	0.4000	1.6000	1.8000	1.8000	II
(2, 1)	0.6000	1.8857	1.8000	1.8857	I
(3, 0)	0.8000	2.4019	1.8000	2.4019	I

K= 3

STATE	R	H(I)	H(II)	EGAIN	S(X,1)
(0, 2)	0.2500	2.0500	2.4000	2.4000	II
(1, 1)	0.5000	2.3429	2.4000	2.4000	II
(2, 0)	0.7500	3.0229	2.4000	3.0229	I

K= 2

STATE	R	H(I)	H(II)	EGAIN	S(X,1)
(0, 1)	0.3333	2.7333	3.0000	3.0000	II
(1, 0)	0.6667	3.4819	3.0000	3.4819	I

K= 1

STATE	R	H(I)	H(II)	EGAIN	S(X,1)
(0, 0)	0.5000	3.7410	3.6000	3.7410	I

Appendix D

ANALYSIS OF A TAB PROBLEM WITH BOTH MACHINES UNKNOWN

PARAMETERS.   N = 5. Machines I, II. R is the unknown Bernoulli parameter of I and S of II. R and S are both described *a priori* by the uniform
density on [0, 1].

ANALYSIS.   The state-space Ω is defined:

$$\Omega = \{(m_1, n_1, m_2, n_2): 0 \leqslant m_j \leqslant n_j, j = 1, 2; n_1 + n_2 \leqslant 4\}. \tag{1}$$

(m_1, n_1, m_2, n_2) denotes that machine I has paid off m_1 times in n_1 trials and
II, m_2 times in n_2 trials.

For all state-times, the control set P is {I, II}. The notation of Chapter 3 is
used extensively. In particular, S_k denotes a strategy constructed according to
dynamic programming. $E[G(x, S_k, k)]$, the expected gain when the system
is in state x at time k and strategy S_k is employed, is defined in the same
fashion as the quantity $E[L(x, S_k, k)]$ in Chapter 3.

Given state x, r(x) and s(x) are the Bayes estimators for R and S. Then,
if $x = (m_1, n_1, m_2, n_2)$,

$$r(x) = (m_1 + 1)/(n_1 + 2),$$
$$s(x) = (m_2 + 1)/(n_2 + 2). \tag{2}$$

For t = 5,

$$H(x, I, t) = r(x) \qquad \text{and} \qquad H(x, II, t) = s(x). \tag{3}$$

If $1 \leqslant t < 5$ and $x = (m_1, n_1, m_2, n_2)$,

$$H(x, I, t) = r(x)\big(1 + E[G((m_1 + 1, n_1 + 1, m_2, n_2), S_{t+1}, t + 1)]\big)$$
$$+ \big(1 - r(x)\big)E[G((m_1, n_1 + 1, m_2, n_2), S_{t+1}, t + 1)] \tag{4}$$

$$H(x, II, t) = s(x)\big(1 + E[G((m_1, n_1, m_2 + 1, n_2 + 1), S_{t+1}, t + 1)]\big)$$
$$+ \big(1 - s(x)\big)E[G((m_1, n_1, m_2, n_2 + 1), S_{t+1}, t + 1)].$$

Also

$$S_t(x, t) = p^*(t), \tag{5}$$

where p*(t) is the element I or II such that

$$H\big(x, p^*(t), t\big) = \min_{p \in \{I, II\}} H(x, p, t). \tag{6}$$

From the symmetry of the problem, the first control element may be
selected arbitrarily.

Appendix E is a computer program and readout for the preceding computations. Some reduction in the state enumeration has been obtained by
observing that, at time t, a state (a, b, c, d) can occur only if $b + d = t - 1$.
In the readout, EGAIN stands for $E[G(x, S, t)]$, where S is a solution.

COMPUTER PROGRAM AND READOUT
FOR TAB PROBLEM,
BOTH MACHINES UNKNOWN

```
*     COMPILE FORTRAN,EXECUTE FORTRAN
*** MAIN PROGRAM ***
    I DEFINED BUT NOT USED IN AN ARITH STMNT.
   II DEFINED BUT NOT USED IN AN ARITH STMNT.
  LNS DEFINED BUT NOT USED IN AN ARITH STMNT.
            DIMENSION LABEL(1600,2),EGAIN(1600,2)
2002      1 FORMAT(I2,2(A2))
2006      2 FORMAT(1H132H*** TWO-ARMED BANDIT PROBLEM ***/5X20HBOTH BANDITS UN
            1KNOWN/5X24HTOTAL NUMBER OF PLAYS = I2)
2026      3 FORMAT(1H13HK= I2//5X5HSTATE12X1HR1OX1HS8X4HH(I)7X5HH(II)6X5HEGAIN
            14X6HS(X,1)/)
2042      4 FORMAT(1X1H(I2,1H,I2,1H,I2,1H,I2,1H)5(5XF6.4),4X2(1XA2))
2054      5 FORMAT(/)


2054   1000 READ 1,220,NTIME,I,II
2063        PRINT 2,NTIME
2068        NC=1

'DO 210' CONTROLS THE TIME PERIOD.
2070        DO 210 IN=1,NTIME
2074        NS=0
2076        LINE=0
2078        NPAGE=0
2080        NN=NTIME-IN
2083        K=NN+1
2086        PRINT 3,K

'DO 180' (NEST OF THREE) GENERATES THE STATE SET (M1,N1,M2,N2).
2091        DO 180 INN=0,NN
2095        N1=NN-INN
2098        N2=K-(N1+1)
2104        DO 180 M2=0,N2
2108        DO 180 M1=0,N1

            COMPUTE R,S,THE ESTIMATORS FOR MACHINES I,II RESPECTIVELY.
2112        RN=M1+1
2116        RD=N1+2
2120        R=RN/RD
2124        SN=M2+1
2128        SD=N2+2
2132        S=SN/SD
2136        LINE=LINE+1
2139        IF(LINE-11)20,8,20
2143      8 LINE=1
2145        NPAGE=NPAGE+1
2148        IF(NPAGE-5)10,9,10
2152      9 NPAGE=0
2154        PRINT 3,K
2159        GO TO 20
2160     10 PRINT 5
2163     20 NS=NS+1
2166        LABEL(NS,NC)=N2+M2*100+N1*10000+M1*1000000
2182        IF(IN-1)35,30,35

            COMPUTE EXPECTED GAIN IN FINAL DECISION PERIOD.
2186     30 H1=R
2188        H2=S
2190        GO TO 140

            COMPUTE EXPECTED GAIN IF MACHINE I IS PLAYED.
2191     35 ICON=N2+M2*100+(N1+1)*10000
2202        DO 80 JJ=1,2
2205        LAST=ICON+(M1-1+JJ)*1000000
2212        DO 40 KK=1,LNS
2216        IF(LAST-LABEL(KK,LNC))40,50,40
2223     40 CONTINUE
2224     50 GO TO (60,70),JJ
2229     60 L=KK
2231        GO TO 80
2232     70 H1=R*(1.+EGAIN(KK,LNC))+(1.-R)*EGAIN(L,LNC)
2253     80 CONTINUE
```

106

```
        COMPUTE EXPECTED GAIN IF MACHINE II IS PLAYED.
2254          ICON=(N2+1)+N1*10000+M1*1000000
2265          DO 130 JJ=1,2
2268          LAST=ICON+(M2-1+JJ)*100
2275          DO 90 KK=1,LNS
2279          IF(LAST-LABEL(KK,LNC))90,100,90
2286       90 CONTINUE
2287      100 GO TO (110,120),JJ
2292      110 L=KK
2294          GO TO 130
2295      120 H2=S*(1.+EGAIN(KK,LNC))+(1.-S)*EGAIN(L,LNC)
2316      130 CONTINUE

        DETERMINE DECISION,S(X,T).
2317      140 IF(H1-H2)150,160,170
2322      150 EGAIN(NS,NC)=H2
2328          PRINT 4,M1,N1,M2,N2,R,S,H1,H2,EGAIN(NS,NC),II
2344          GO TO 180
2345      160 EGAIN(NS,NC)=H1
2351          PRINT 4,M1,N1,M2,N2,R,S,H1,H2,EGAIN(NS,NC),I,II
2368          GO TO 180
2369      170 EGAIN(NS,NC)=H1
2375          PRINT 4,M1,N1,M2,N2,R,S,H1,H2,EGAIN(NS,NC),I
2391      180 CONTINUE
2394          LNS=NS
2396          GO TO (190,200),NC
2401      190 NC=2
2403          LNC=1
2405          GO TO 210
2406      200 NC=1
2408          LNC=2
2410      210 CONTINUE
2411          GO TO 1000
2412      220 STOP
              END

        MAIN PROGRAM
```

*** TWO-ARMED BANDIT PROBLEM ***
BOTH BANDITS UNKNOWN
TOTAL NUMBER OF PLAYS = 5

K= 5

STATE	R	S	H(I)	H(II)	EGAIN	S(X,1)
(0, 4, 0, 0)	0.1667	0.5000	0.1667	0.5000	0.5000	II
(1, 4, 0, 0)	0.3333	0.5000	0.3333	0.5000	0.5000	II
(2, 4, 0, 0)	0.5000	0.5000	0.5000	0.5000	0.5000	I II
(3, 4, 0, 0)	0.6667	0.5000	0.6667	0.5000	0.6667	I
(4, 4, 0, 0)	0.8333	0.5000	0.8333	0.5000	0.8333	I
(0, 3, 0, 1)	0.2000	0.3333	0.2000	0.3333	0.3333	II
(1, 3, 0, 1)	0.4000	0.3333	0.4000	0.3333	0.4000	I
(2, 3, 0, 1)	0.6000	0.3333	0.6000	0.3333	0.6000	I
(3, 3, 0, 1)	0.8000	0.3333	0.8000	0.3333	0.8000	I
(0, 3, 1, 1)	0.2000	0.6667	0.2000	0.6667	0.6667	II
(1, 3, 1, 1)	0.4000	0.6667	0.4000	0.6667	0.6667	II
(2, 3, 1, 1)	0.6000	0.6667	0.6000	0.6667	0.6667	II
(3, 3, 1, 1)	0.8000	0.6667	0.8000	0.6667	0.8000	I
(0, 2, 0, 2)	0.2500	0.2500	0.2500	0.2500	0.2500	I II
(1, 2, 0, 2)	0.5000	0.2500	0.5000	0.2500	0.5000	I
(2, 2, 0, 2)	0.7500	0.2500	0.7500	0.2500	0.7500	I
(0, 2, 1, 2)	0.2500	0.5000	0.2500	0.5000	0.5000	II
(1, 2, 1, 2)	0.5000	0.5000	0.5000	0.5000	0.5000	I II
(2, 2, 1, 2)	0.7500	0.5000	0.7500	0.5000	0.7500	I
(0, 2, 2, 2)	0.2500	0.7500	0.2500	0.7500	0.7500	II

K= 5

STATE	R	S	H(I)	H(II)	EGAIN	S(X,1)
(1, 2, 2, 2)	0.5000	0.7500	0.5000	0.7500	0.7500	II
(2, 2, 2, 2)	0.7500	0.7500	0.7500	0.7500	0.7500	I II
(0, 1, 0, 3)	0.3333	0.2000	0.3333	0.2000	0.3333	I
(1, 1, 0, 3)	0.6667	0.2000	0.6667	0.2000	0.6667	I
(0, 1, 1, 3)	0.3333	0.4000	0.3333	0.4000	0.4000	II
(1, 1, 1, 3)	0.6667	0.4000	0.6667	0.4000	0.6667	I
(0, 1, 2, 3)	0.3333	0.6000	0.3333	0.6000	0.6000	II
(1, 1, 2, 3)	0.6667	0.6000	0.6667	0.6000	0.6667	I
(0, 1, 3, 3)	0.3333	0.8000	0.3333	0.8000	0.8000	II
(1, 1, 3, 3)	0.6667	0.8000	0.6667	0.8000	0.8000	II
(0, 0, 0, 4)	0.5000	0.1667	0.5000	0.1667	0.5000	I
(0, 0, 1, 4)	0.5000	0.3333	0.5000	0.3333	0.5000	I
(0, 0, 2, 4)	0.5000	0.5000	0.5000	0.5000	0.5000	I II
(0, 0, 3, 4)	0.5000	0.6667	0.5000	0.6667	0.6667	II
(0, 0, 4, 4)	0.5000	0.8333	0.5000	0.8333	0.8333	II

K= 4

STATE	R	S	H(I)	H(II)	EGAIN	S(X,1)
(0, 3, 0, 0)	0.2000	0.5000	0.7000	1.0000	1.0000	II
(1, 3, 0, 0)	0.4000	0.5000	0.9000	1.0333	1.0333	II
(2, 3, 0, 0)	0.6000	0.5000	1.2000	1.1333	1.2000	I
(3, 3, 0, 0)	0.8000	0.5000	1.6000	1.3000	1.6000	I
(0, 2, 0, 1)	0.2500	0.3333	0.6000	0.6667	0.6667	II
(1, 2, 0, 1)	0.5000	0.3333	1.0000	0.8333	1.0000	I
(2, 2, 0, 1)	0.7500	0.3333	1.5000	1.0833	1.5000	I
(0, 2, 1, 1)	0.2500	0.6667	0.9167	1.3333	1.3333	II
(1, 2, 1, 1)	0.5000	0.6667	1.1667	1.3333	1.3333	II
(2, 2, 1, 1)	0.7500	0.6667	1.5167	1.4167	1.5167	I
(0, 1, 0, 2)	0.3333	0.2500	0.6667	0.6000	0.6667	I
(1, 1, 0, 2)	0.6667	0.2500	1.3333	0.9167	1.3333	I
(0, 1, 1, 2)	0.3333	0.5000	0.8333	1.0000	1.0000	II
(1, 1, 1, 2)	0.6667	0.5000	1.3333	1.1667	1.3333	I
(0, 1, 2, 2)	0.3333	0.7500	1.0833	1.5000	1.5000	II
(1, 1, 2, 2)	0.6667	0.7500	1.4167	1.5167	1.5167	II
(0, 0, 0, 3)	0.5000	0.2000	1.0000	0.7000	1.0000	I
(0, 0, 1, 3)	0.5000	0.4000	1.0333	0.9000	1.0333	I
(0, 0, 2, 3)	0.5000	0.6000	1.1333	1.2000	1.2000	II
(0, 0, 3, 3)	0.5000	0.8000	1.3000	1.6000	1.6000	II

K= 3

STATE	R	S	H(I)	H(II)	EGAIN	S(X,1)
(0, 2, 0, 0)	0.2500	0.5000	1.2583	1.5000	1.5000	II
(1, 2, 0, 0)	0.5000	0.5000	1.6167	1.6667	1.6667	II
(2, 2, 0, 0)	0.7500	0.5000	2.2500	2.0083	2.2500	I
(0, 1, 0, 1)	0.3333	0.3333	1.1111	1.1111	1.1111	I II
(1, 1, 0, 1)	0.6667	0.3333	2.0000	1.6667	2.0000	I
(0, 1, 1, 1)	0.3333	0.6667	1.6667	2.0000	2.0000	II
(1, 1, 1, 1)	0.6667	0.6667	2.1222	2.1222	2.1222	I II
(0, 0, 0, 2)	0.5000	0.2500	1.5000	1.2583	1.5000	I
(0, 0, 1, 2)	0.5000	0.5000	1.6667	1.6167	1.6667	I
(0, 0, 2, 2)	0.5000	0.7500	2.0083	2.2500	2.2500	II

K= 2

STATE	R	S	H(I)	H(II)	EGAIN	S(X,1)
(0, 1, 0, 0)	0.3333	0.5000	1.8889	2.0556	2.0556	II
(1, 1, 0, 0)	0.6667	0.5000	2.7222	2.5611	2.7222	I
(0, 0, 0, 1)	0.5000	0.3333	2.0556	1.8889	2.0556	I
(0, 0, 1, 1)	0.5000	0.6667	2.5611	2.7222	2.7222	II

K= 1

STATE	R	S	H(I)	H(II)	EGAIN	S(X,1)
(0, 0, 0, 0)	0.5000	0.5000	2.8889	2.8889	2.8889	I II

THEOREM FOR CONSTRUCTING SOLUTIONS FOR DISCOUNTED TABS

THEOREM 4.2.   *Assume* s, *the Bernoulli parameter of machine II, is chosen by a random mechanism whose probability distribution has no mass points. Then with probability 1, for each number* n, *there is some positive integer* N *such that*

$$E[G_N((u_N(n), n), S_N, 1)] - E[G_N((u_N(n), n), S_N'', 1)] > a^N/(1 - a).$$

PROOF.   Suppose, for all N,

$$E[G((u_N(n), n), S_N, 1)] - E[G_N((u_N(n), n), S_N'', 1)] \leq a^N/(1 - a). \quad (1)$$

The argument leading to (4.23) implies that $u_N(n)$ is monotonically decreasing in N. Let

$$m = \min_N u_N(n) \quad \text{and} \quad x = (m, n).$$

Then $E[G(x, S, 1)] = s/(1 - a)$ and thus $m = u(n)$. Define S'' to be a best strategy such that I is always used at the first play. Then

$$E[G(x, S, 1)] = E[G(x, S'', 1)]. \quad (2)$$

For $S(x, 1) = II$ and therefore $E[G(x, S'', 1)] \leqslant E[G(x, S, 1)]$. Suppose

$$E[G(x, S'', 1)] = E[G(x, S, 1)] - c,$$

where $c > 0$. Pick q so that $a^q/(1 - a) < c/2$ and $u_q(n) = m$. Then

$$\begin{aligned}
E[G_q(x, S_q'', 1)] &\geqslant E[G_q(x, S_q, 1)] - a^q/(1 - a) \\
&= s(1 - a^q)/(1 - a) - a^q/(1 - a) \\
&> s/(1 - a) - c,
\end{aligned}$$

which would imply that S_q'' extended arbitrarily to decision times greater than q is better than S'', contrary to the definition of S''.

Suppose (1) is true for all N. Then in view of (2),

$$\begin{aligned}
E[G((m, n), S, 1)] &= s/(1 - a) \\
&= r(m, n)(1 + E[G((m + 1, n), S'', 2)]) \\
&\quad + (1 - r(m, n))E[G((m, n + 1), S'', 2)]
\end{aligned}$$

If

$$E[G((m + 1, n), S'', 2)] \neq E[G((m + 1, n), S, 2)]$$

then S'' would be dominated by the strategy which plays I at time 1 and agrees with S at other times. Similarly $E[G((m, n + 1), S'', 2)]$ must be equal to $E[G((m, n + 1), S, 2)]$.

From the symmetry of the infinite discounted bandit,

$$E[G(x, S, t)] = a^{t-1}E[G(x, S, 1)].$$

Also, since $s(m, n) > s(m, n + 1)$ if $E[G((m, n), S, 1)] = s/(1 - a)$, then $E[G((m, n + 1), S, 1)] = s/(1 - a)$. In summary,

$s/(1 - a)$

$$= r(m, n)(1 + aE[G((m + 1, n), S, 1)]) + (1 - r(m, n))as/(1 - a).$$

Explicitly in terms of $E[G((m + 1, n), S, 1)]$ this becomes

$E[G((m + 1, n), S, 1)]$

$$= ([1/ar(m, n)][s/(1 - a)](1 - a + ar(m, n))) - 1/a.$$

Therefore, if (1) holds for two values, $s_1 > s_2$, of s, and S_{s_1} and S_{s_2} are Theorem 4.1 solutions for the indicated values of s, and $E_s[\cdot]$ denotes the expected value with respect to the indicated value of the parameter s,

$$E_{s_1}[G((m + 1, n), S_{s_1}, 1)] - E_{s_2}[G((m + 1, n), S_{s_2}, 1)]$$

$$= 1/ar(m, n)[(s_1 - s_2)/(1 - a)](1 - a + ar(m, n)) > (s_1 - s_2)/(1 - a). \quad (3)$$

But always,

$$E_{s_1}[G((m + 1, n), S_{s_1}, 1)] - E_{s_2}[G((m + 1, n), S_{s_2}, 1)] \leqslant (s_1 - s_2)/(1 - a).$$

For suppose $\{u(j)\}$ is the sequence associated with S_{s_1} and $p(j)$ is the probability that state $(u(j), j)$ is reached under strategy S_{s_1} from initial state $(m + 1, n)$. Let $G(j)$ be the expected return in reaching $(u(j), j)$ and $E[I]$ the expected return given none of the states $\{(u(j), j)\}$ is reached. Note that $p(j)$, $G(j)$, and $E[I]$ do not depend on the parameter s.

For any s,

$$E_s[G((m + 1, n), S_{s_1}, 1)] = \sum_{j=1}^{\infty} p(j)[(G(j) + sa^j/(1 - a))] + E[I]\left(1 - \sum_{j=1}^{\infty} p(j)\right),$$

and in particular

$$E_{s_1}[G((m + 1, n), S_{s_1}, 1)] - E_{s_2}[G((m + 1, n), S_{s_2}, 1)]$$

$$\leq E_{s_1}[G((m + 1, n), S_{s_1}, 1)] - E_{s_2}[G((m + 1, n), S_{s_1}, 1)]$$

$$= \left(\sum_{j=1}^{\infty} p(j)a^j\right)[(s_1 - s_2)/(1 - a)] < \sum_{j=1}^{\infty} p(j)[(s_1 - s_2)/(1 - a)]$$

$$\leq (s_1 - s_2)/(1 - a).$$

Now it can be concluded that (1) cannot hold for two values of s, and if the probability law choosing s has no mass points the value at which (1) is true occurs with probability 0.

Appendix G

COMPUTER STUDIES ON AN INFINITE-PLAY BANDIT

```
*       COMPILE FORTRAN,EXECUTE FORTRAN
*** MAIN PROGRAM ***
NEWV DEFINED BUT NOT USED IN AN ARITH STMNT.
MAXV DEFINED BUT NOT USED IN AN ARITH STMNT.
             DIMENSION EGAIN(1000,2),HI(100),HII(100)
2002     10 FORMAT(I2,8X4(I4,6X),2F10.4)
2008     20 FORMAT(1H132H*** TWO ARMED BANDIT PROBLEM ***/5X44HINITIAL SEQUENC
            1E TO DISCOUNTED INFINITE GAME/5X28HBANDIT NUMBER TWO KNOWN, S= F6.
            24/5X20HDISCOUNT FACTOR, A= F6.4)
2041     30 FORMAT(1H11X9H(U*(V),V)13X2HGN17X3HGN'16X6HGN-GN'12X10HA**N/(1-A)
            19X1HN/)
2056     40 FORMAT( 1X1H(I4,1H,I4,1H)4E20.8,5X14)
2064     50 FORMAT(1H15X30HINITIAL SEQUENCE IS OF LENGTH I4//6X60HTO CONTINUE
            1SEQUENCE,ENTER THE FOLLOWING VALUES ON DATA CARD/10X98HNSTAR...REM
            2AINS SAME,OR CAN INCREASE DEPENDING ON TREND OF 'N' NEAR END OF SE
            3QUENCE JUST GENERATED/10X6HLASTU=I4/10X6HNEWV =I4/10X36HMAXV...CHO
            4OSE A TERMINAL VALUE FOR V/10X6HLONG =I4/10X6HS...REMAINS SAME/
            510X16HR...REMAINS SAME)
2134     60 FORMAT(/)


2135        READ 10,NSTAR,LASTU,NEWV,MAXV,LONG,S,A
2147        PRINT 20,S,A
2153        IU=LASTU
2155        LINE=0
2157        NPAGE=0
2159        PRINT 30

'DO 290' ALLOWS 'V' TO RANGE THROUGH DESIRED VALUES.
2162        DO 290 IV=NEWV,MAXV
2167        NTIME=NSTAR
2169    100 H2=0
2172        KK=0
2174        NC=1
2176        BOUND=A**NTIME/(1.-A)

'DO 200' INCREMENTS 'J' AS REQUIRED TO DETERMINE UN*(V).
2185        DO 200 JJ=IU,IV
2190        KK=KK+1

    'DO 190' GENERATES THE ELEMENT 'LOWER CASE N' IN THE STATE SET

    OF A GAME OF LENGTH 'CAPITAL N'.
2193        DO 190 IN=1,NTIME
2197        N=NTIME-IN

    COMPUTE H2 AND DISCOUNT FOR PRESENT DECISION PERIOD.
2200        DISC=A**N
2204        H2=DISC*S+H2
2208        HII(KK)=H2
2211        NS=0

    'DO 160' GENERATES THE ELEMENT 'LOWER CASE M' IN STATE SET.
2213        DO 160 M=0,N
2217        NS=NS+1

    COMPUTE 'R',THE ESTIMATOR FOR MACHINE I.
2220        RN=M+JJ+1
2225        RD=N+IV+2
2230        R=RN/RD

    AT FINAL DECISION TIME H1 IS COMPUTED AS PER STMNT. 110,

    OTHERWISE AS PER STMNT. 120.
2234        IF(IN-1)120,110,120
2238    110 H1=R*DISC
2241        GO TO 130
2242    120 H1=R*(DISC+EGAIN(NS+1,LNC))+(1.-R)*EGAIN(NS,LNC)
2259    130 DIFF=H2-H1
2262        HI(KK)=H1
```

```
      DETERMINE DECISION SN(X,T).
2265         IF(DIFF)150,140,140
2268     140 EGAIN(NS,NC)=H2
2274         GO TO 160
2275     150 EGAIN(NS,NC)=H1
2281     160 CONTINUE
2282         GO TO(170,180),NC
2287     170 NC=2
2289         LNC=1
2291         GO TO 190
2292     180 NC=1
2294         LNC=2
2296     190 CONTINUE

   SEARCH FOR REQUIRED 'J' TERMINATES THE FIRST TIME SN((JJ,V),1)=I

   THEN J=JJ-1 AND UN*(V)=J.
2297         IF(DIFF)210,200,200
2300     200 CONTINUE

   IF FOLLOWING STMNT. IS EVER REACHED,THE SOLUTION IS ALWAYS PLAY

   MACHINE II.
2301         J=JJ
2303         LONG=LONG+1
2306         PRINT 40,J,IV,H2,H1,DIFF,BOUND,NTIME
2317         GO TO 300

   BRANCH TO STMNT. 220 IS TAKEN ONLY IF SN((U,V),1)=I,

   THIS SHOULD ONLY OCCUR FOR (U,V)=(0,0).
2318     210 IF(JJ-IU)230,220,230
2322     220 J=JJ-1
2325         GO TO 260
2326     230 DIFF=HII(KK-1)-HI(KK-1)

   IF GN-GN' .GT. A**N/(1-A),THEN U*(V)=UN*(V).

   IF GN-GN' .LE. A**N/(1-A),THEN UN*(V) .GT. U*(V),


   INCREASE N AND TRY AGAIN.
2330         IF(DIFF-BOUND)240,240,250
2335     240 NTIME=NTIME+1
2338         GO TO 100
2339     250 J=JJ-1
2342         IU=J
2344         H2=HII(KK-1)
2347         H1=HI(KK-1)
2349     260 LONG=LONG+1
2352         LINE=LINE+1
2355         IF(LINE-11)286,270,286
2359     270 LINE=1
2361         NPAGE=NPAGE+1
2364         IF(NPAGE-5)285,280,285
2368     280 NPAGE=0
2370         PRINT 30
2373         GO TO 286
2374     285 PRINT 60
2377     286 PRINT 40,J,IV,H2,H1,DIFF,BOUND,NTIME
2388     290 CONTINUE
2389     300 NEWV=IV+1

   TERMINATE SEQUENCE,PRINT NEW STARTING INFO. TO CONTINUE SEQUENCE
2392         PRINT 50,LONG,IU,NEWV,LONG
2400         STOP
             END

MAIN PROGRAM
```

117

*** TWO ARMED BANDIT PROBLEM ***
INITIAL SEQUENCE TO DISCOUNTED INFINITE GAME
BANDIT NUMBER TWO KNOWN, S= 0.6000
DISCOUNT FACTOR, A= 0.8000

(U*(V),V)	GN	GN'	GN-GN'	A**N/(1-A)	N
(-1, 0)	0.29654123E 01	0.30588237E 01	-0.93411400E-01	0.57646070E-01	20
(0, 1)	0.29654123E 01	0.26987455E 01	0.26666680E 00	0.57646070E-01	20
(1, 2)	0.59773317E 01	0.59540376E 01	0.23294100E-01	0.18889464E-01	25
(1, 3)	0.29654123E 01	0.27654122E 01	0.20000010E 00	0.57646070E-01	20
(2, 4)	0.59308240E 01	0.58568816E 01	0.73942400E-01	0.57646070E-01	20
(2, 5)	0.29654123E 01	0.27939836E 01	0.17142870E 00	0.57646070E-01	20
(3, 6)	0.59308240E 01	0.58320046E 01	0.98819400E-01	0.57646070E-01	20
(3, 7)	0.29654123E 01	0.28098566E 01	0.15555570E 00	0.57646070E-01	20
(4, 8)	0.59308240E 01	0.58308240E 01	0.10000000E 00	0.57646070E-01	20
(5, 9)	0.59645814E 01	0.59302231E 01	0.34358300E-01	0.29514789E-01	23
(5, 10)	0.29654123E 01	0.28654122E 01	0.10000010E 00	0.57646070E-01	20
(6, 11)	0.59308240E 01	0.58717261E 01	0.59097900E-01	0.57646070E-01	20
(6, 12)	0.29654123E 01	0.28654122E 01	0.10000010E 00	0.57646070E-01	20
(7, 13)	0.59308240E 01	0.58641572E 01	0.66666800E-01	0.57646070E-01	20
(8, 14)	0.59773317E 01	0.59558463E 01	0.21485400E-01	0.18889464E-01	25
(8, 15)	0.29654123E 01	0.28948240E 01	0.70588300E-01	0.57646070E-01	20
(9, 16)	0.59557270E 01	0.59139897E 01	0.41737300E-01	0.36893487E-01	22
(9, 17)	0.29654123E 01	0.28917279E 01	0.73684400E-01	0.57646070E-01	20
(10, 18)	0.59446590E 01	0.58946589E 01	0.50000100E-01	0.46116859E-01	21
(11, 19)	0.59818652E 01	0.59664750E 01	0.15390200E-01	0.15111572E-01	26

(U*(V),V)	GN	GN'	GN—GN'	A**N/(1—A)	N
(11, 20)	0.29723298E 01	0.29177842E 01	0.54545600E-01	0.46116859E-01	21
(12, 21)	0.59645814E 01	0.59323883E 01	0.32193100E-01	0.29514789E-01	23
(12, 22)	0.29654123E 01	0.29070789E 01	0.58333400E-01	0.57646070E-01	20
(13, 23)	0.59557270E 01	0.59157269E 01	0.40000100E-01	0.36893487E-01	22
(14, 24)	0.59883933E 01	0.59765279E 01	0.11865400E-01	0.96714060E-02	28
(14, 25)	0.29778638E 01	0.29341929E 01	0.44444600E-01	0.36893487E-01	22
(15, 26)	0.59716649E 01	0.59455053E 01	0.26159600E-01	0.23611831E-01	24
(15, 27)	0.29723298E 01	0.29240539E 01	0.48275900E-01	0.46116859E-01	21
(16, 28)	0.59645814E 01	0.59312479E 01	0.33333500E-01	0.29514789E-01	23
(17, 29)	0.59907143E 01	0.59810914E 01	0.96229000E-02	0.77371250E-02	29
(17, 30)	0.29778638E 01	0.29403637E 01	0.37500100E-01	0.36893487E-01	22
(18, 31)	0.59773317E 01	0.59553207E 01	0.22011000E-01	0.18889464E-01	25
(18, 32)	0.29778638E 01	0.29366872E 01	0.41176600E-01	0.36893487E-01	22
(19, 33)	0.59716649E 01	0.59430934E 01	0.28571500E-01	0.23611831E-01	24
(20, 34)	0.59907143E 01	0.59826320E 01	0.80823000E-02	0.77371250E-02	29
(20, 35)	0.29822910E 01	0.29498585E 01	0.32432500E-01	0.29514789E-01	23
(21, 36)	0.59773317E 01	0.59583254E 01	0.19006300E-01	0.18889464E-01	25
(21, 37)	0.29822910E 01	0.29463935E 01	0.35897500E-01	0.29514789E-01	23
(22, 38)	0.59716649E 01	0.59466648E 01	0.25000100E-01	0.23611831E-01	24
(23, 39)	0.59925712E 01	0.59856251E 01	0.69461000E-02	0.61897000E-02	30
(23, 40)	0.29858328E 01	0.29572612E 01	0.28571600E-01	0.23611831E-01	24
(24, 41)	0.59818652E 01	0.59651622E 01	0.16703000E-01	0.15111572E-01	26
(24, 42)	0.29822910E 01	0.29504727E 01	0.31818300E-01	0.29514789E-01	23
(25, 43)	0.59773317E 01	0.59551093E 01	0.22224400E-01	0.18889464E-01	25
(26, 44)	0.59940566E 01	0.59879738E 01	0.60828000E-02	0.49517600E-02	31
(26, 45)	0.29858328E 01	0.29603008E 01	0.25532000E-01	0.23611831E-01	24
(27, 46)	0.59854918E 01	0.59705974E 01	0.14894400E-01	0.12089257E-01	27
(27, 47)	0.29858328E 01	0.29572612E 01	0.28571600E-01	0.23611831E-01	24
(28, 48)	0.59773317E 01	0.59573316E 01	0.20000100E-01	0.18889464E-01	25
(29, 49)	0.59940566E 01	0.59886450E 01	0.54116000E-02	0.49517600E-02	31

(U*(V),V)	GN	GN'	GN-GN'	A**N/(1-A)	N
(29, 50)	0.29886662E 01	0.29655892E 01	0.23077000E-01	0.18889464E-01	25
(30, 51)	0.59854918E 01	0.59720453E 01	0.13446500E-01	0.12089257E-01	27
(30, 52)	0.29858328E 01	0.29590068E 01	0.25926000E-01	0.23611831E-01	24
(31, 53)	0.59818652E 01	0.59636832E 01	0.18182000E-01	0.15111572E-01	26
(32, 54)	0.59952450E 01	0.59903787E 01	0.48863000E-02	0.39614079E-02	32
(32, 55)	0.29886662E 01	0.29676135E 01	0.21052700E-01	0.18889464E-01	25
(33, 56)	0.59854918E 01	0.59732375E 01	0.12254300E-01	0.12089257E-01	27
(33, 57)	0.29858328E 01	0.29621039E 01	0.23728900E-01	0.23611831E-01	24
(34, 58)	0.59818652E 01	0.59651984E 01	0.16666800E-01	0.15111572E-01	26
(35, 59)	0.59952450E 01	0.59908221E 01	0.44229000E-02	0.39614079E-02	32
(35, 60)	0.29886662E 01	0.29693113E 01	0.19354900E-01	0.18889464E-01	25
(36, 61)	0.59883933E 01	0.59714455E 01	0.11247800E-01	0.96714060E-02	28
(36, 62)	0.29886662E 01	0.29667911E 01	0.21875100E-01	0.18889464E-01	25
(37, 63)	0.59818652E 01	0.59664805E 01	0.15384700E-01	0.15111572E-01	26
(38, 64)	0.59952450E 01	0.59911923E 01	0.40527000E-02	0.39614079E-02	32
(38, 65)	0.29909330E 01	0.29730224E 01	0.17910600E-01	0.15111572E-01	26
(39, 66)	0.59883933E 01	0.59799935E 01	0.10399800E-01	0.96714060E-02	28
(39, 67)	0.29886662E 01	0.29683762E 01	0.20290000E-01	0.18889464E-01	25
(40, 68)	0.59854918E 01	0.59712060E 01	0.14285800E-01	0.12089257E-01	27
(41, 69)	0.59961957E 01	0.59924604E 01	0.37353000E-02	0.31691263E-02	33
(41, 70)	0.29909330E 01	0.29742662E 01	0.16666800E-01	0.15111572E-01	26
(42, 71)	0.59907143E 01	0.59810493E 01	0.96650000E-02	0.77371250E-02	29
(42, 72)	0.29886662E 01	0.29697472E 01	0.18919000E-01	0.18889464E-01	25
(43, 73)	0.59854918E 01	0.59721584E 01	0.13333400E-01	0.12089257E-01	27
(44, 74)	0.59961957E 01	0.59927295E 01	0.34662000E-02	0.31691263E-02	33
(44, 75)	0.29909330E 01	0.29753485E 01	0.15584500E-01	0.15111572E-01	26
(45, 76)	0.59907143E 01	0.59816831E 01	0.90312000E-02	0.77371250E-02	29
(45, 77)	0.29909330E 01	0.29732114E 01	0.17721600E-01	0.15111572E-01	29
(46, 78)	0.59854918E 01	0.59729917E 01	0.12500100E-01	0.12089257E-01	26
(47, 79)	0.59961957E 01	0.59929630E 01	0.32327000E-02	0.31691263E-02	33

(U*(V),V)		GN	GN'	GN-GN'	A**N/(1-A)	N
(47,	80)	0.29927463E 01	0.29781121E 01	0.14634200E-01	0.12089257E-01	27
(48,	81)	0.59907143E 01	0.59823389E 01	0.84754000E-02	0.77371250E-02	29
(48,	82)	0.29909330E 01	0.29742662E 01	0.16666800E-01	0.15111572E-01	26
(49,	83)	0.59883933E 01	0.59766285E 01	0.11764800E-01	0.96714060E-02	28
(50,	84)	0.59969562E 01	0.59939298E 01	0.30264000E-02	0.25353010E-02	34
(50,	85)	0.29927463E 01	0.29789531E 01	0.13793200E-01	0.12089257E-01	27
(51,	86)	0.59907143E 01	0.59827305E 01	0.79838000E-02	0.77371250E-02	29
(51,	87)	0.29909330E 01	0.29752025E 01	0.15730500E-01	0.15111572E-01	26
(52,	88)	0.59883933E 01	0.59772821E 01	0.11111200E-01	0.96714060E-02	28
(53,	89)	0.59969562E 01	0.59941100E 01	0.28462000E-02	0.25353010E-02	34
(53,	90)	0.29927463E 01	0.29797027E 01	0.13043600E-01	0.12089257E-01	27
(54,	91)	0.59925712E 01	0.59850287E 01	0.75425000E-02	0.61897000E-02	30
(54,	92)	0.29927463E 01	0.29778526E 01	0.14893700E-01	0.12089257E-01	27
(55,	93)	0.59883933E 01	0.59778669E 01	0.10526400E-01	0.96714060E-02	28
(56,	94)	0.59969562E 01	0.59942700E 01	0.26862000E-02	0.25353010E-02	34
(56,	95)	0.29927463E 01	0.29803751E 01	0.12371200E-01	0.12089257E-01	27
(57,	96)	0.59925712E 01	0.59854207E 01	0.71505000E-02	0.61897000E-02	30
(57,	97)	0.29927463E 01	0.29786048E 01	0.14141500E-01	0.12089257E-01	27
(58,	98)	0.59883933E 01	0.59783932E 01	0.10000100E-01	0.96714060E-02	28
(59,	99)	0.59969562E 01	0.59944132E 01	0.25430000E-02	0.25353010E-02	34

INITIAL SEQUENCE IS OF LENGTH 100

TO CONTINUE SEQUENCE,ENTER THE FOLLOWING VALUES ON DATA CARD
NSTAR...REMAINS SAME,OR CAN INCREASE DEPENDING ON TREND OF 'N' NEAR END OF SEQUENCE JUST GENERATED
LASTU= 59
NEWV = 100
MAXV...CHOOSE A TERMINAL VALUE FOR V
LONG = 100
S...REMAINS SAME
R...REMAINS SAME

Appendix H

THEOREM ON THE RATE OF CONVERGENCE OF EMPIRICAL DISTRIBUTION FUNCTIONS

LEMMA. *Let* X_n *be the* n-th *sample mean of a sequence of independent r.v.'s* $\{Y_i\}$, *each identically distributed according to a Bernoulli law with parameter* p, *which may be unknown. Then, given any positive numbers* c_1, c_2, *a number* N *may be computed such that, regardless of the parameter* p,

$$P\left[\sup_{n > N} |X_n - p| > c_1\right] < c_2. \tag{1}$$

PROOF.   It is easily seen that var $(X_n) = p(1 - p)/n \leqslant 1/4n$. Also, $E[X_n] = p$. For some positive integer k, define E to be the event that, if $n \geqslant k$, $|X_{n^2} - p| < c_1/2$. Suppose m is any number greater than k^2. Then there is a number v such that

$$k^2 \leqslant v^2 \leqslant m < (v + 1)^2. \tag{2}$$

If, for all j, $S_j \equiv j(X_j)$,*

$$|X_m - X_{v^2}| = |S_m/m - S_{v^2}/v^2| \leq \max \begin{cases} S_{v^2}/v^2 - S_{v^2}/(v + 1)^2 \equiv \zeta(1), \\ (S_{(v+1)^2-1})/v^2 - S_{v^2}/v^2 \equiv \zeta(2), \end{cases} \tag{3}$$

$$\zeta(2) \leqslant 2v/v^2 = 2/v \leqslant 2/k, \tag{4}$$

$$\zeta(1) = S_{v^2}(1/v^2 - 1/(v + 1)^2) \leqslant v^2(2v/v^2(v + 1)^2) < 2/v \leqslant 2/k. \tag{5}$$

Therefore, if event E happens and also $m > k^2$,

$$|p - X_m| \leqslant |p - X_{v_2}| + |X_{v^2} - X_m| < 2/k + c_1/2. \tag{6}$$

Define E_j to be the event $|X_{j^2} - p| < c_1/2$. By Tchebichev's inequality,

$$P[E_j^c] \leqslant (2/c_1)^2(1/j^2)(1/4) = (1/c_1^2)(1/j)^2. \tag{7}$$

Subadditivity gives

$$P[E^c] = P\left[\bigcup_{j=k}^{\infty} E_j^c\right] \leqslant \sum_{j=k}^{\infty} P[E_j^c] \leqslant (1/c_1)^2\left(\sum_{j=k}^{\infty} 1/j^2\right). \tag{8}$$

Let Q_1 be the smallest integer such that

$$c_2 > \left(\frac{1}{c_1}\right)^2 \sum_{j=Q_1}^{\infty} 1/j^2. \tag{9}$$

Then E^c occurs with probability less than c_2 if k is taken to be Q_1. If Q_2 is the smallest positive integer such that $1/Q_2 < c_1/2$, $Q = \max \{Q_1, Q_2\}$, and event E happens, then for any $n \geqslant Q^2$, from (6),

$$|X_n - p| < 1/Q_2 + c_1/2 < c_1.$$

* Note: The use of S_j here is different from elsewhere in this book.

In summary, if $Q = Q(c_1, c_2)$ is the smallest number satisfying the two conditions

$$c_1^2 c_2 > \sum_{j=Q(c_1,c_2)}^{\infty} 1/j^2 \quad \text{and} \quad 1/Q(c_1, c_2) < c_1/2,$$

then, with probability greater than $1 - c_2$, for each $n > (Q(c_1, c_2))^2$, $|X_n - p| < c_1$. In the following theorem, $N(c_1, c_2) = (Q(c_1, c_2))^2$.

THEOREM. *Assume:*
(1) $F(x) = P[X < x]$, *and hence F is continuous from below.*
(2) $\rho(F_1, F_2) = \sup\limits_{x \in R} | F_1(x) - F_2(x)|$.

Let $\{X_i\}$ be a sequence of independent trials of X, a r.v. distributed according to some abritrary c.d.f. F. Let $\{F_i\}$ be the sequence of functions such that F_n is the empirical distribution function constructed from the first n trials of $\{X_i\}$. Then given any positive numbers c_1, c_2, a number N dependent on c_1 and c_2 but not on F, may be found such that

$$P\left[\sup_{z > N} \rho(F_n, F) > c_1 \right] < c_2. \tag{10}$$

PROOF. Let M be the least positive integer such that $1/M < c_1/8$. $Y = \{y_i\}_{i=1}^{M}$ is defined so that, for $1 \leqslant i \leqslant M$, y_i is the greatest extended real* number such that

$$F(y_i) \leqslant i/M \leqslant F(y_i+),$$

where for all real x, $F(y+) = \inf\limits_{z>x} F(z)$. $y_0 = -\infty$.

Observe that the events "$X < y_i$" and "$X \leqslant y_i$" are Bernoulli trials with parameters $F(y_i)$ and $F(y_i+)$, respectively. Thus if

$$n \geqslant N(c_1/8, c_2/(2(M + 1))) \equiv N,$$

then from the lemma

$$P[|F_n(y_i) - F(y_i)| > c_1/8] < c_2/(2(M + 1)).$$

$E_j, j = 0, 1, \ldots, M$, is defined to be the event that, for all $n \geqslant N$,

$$|F_n(y_i) - F(y_i)| < c_1/8,$$

and similarly, the event D_j implies that $|F_n(y_i+) - F(y_i+)| < c_1/8$. The event

* That is, y_i may be ∞.

E is the joint occurrence of all D_j and E_j, $j = 0, 1, \ldots, M$. Thus

$$P[E^c] = P\left[\bigcup_{j=0}^{M} (E_j^c \cup D_j^c) \right]$$

$$\leq \sum_{j=0}^{M} \left(P[E_j^c] + P(D_j^c) \right)$$

$$= 2(M + 1)\left(c_2/(2(M + 1))\right)$$

$$= c_2.$$

If $x = y_i \in Y$ and event E occurs, then it is trivially true that

$$|F_n(x) - F(x)| < c_1$$

for all $n > N$. Assume $x \in R - Y$ and v is an integer such that $x \in (y_v, y_{v+1})$. Repeated use of the triangle inequality gives

$$|F_n(x) - F(x)| \leqslant |F_n(x) - F_n(y_v+)| + |F_v(y_v+) - F(y_v+)|$$

$$+ |F(y_v+) - F(x)|.$$

As c.d.f's are monotone increasing,

$$F_n(y_{v+1}) \geqslant F_n(x) \geqslant F_n(y_v+)$$

and

$$F(y_{v+1}) \geqslant F(x) \geqslant F(y_v+).$$

Observe that, by the definition of Y,

$$F(y_v+) \geqslant v/M \quad \text{and} \quad F(y_{v+1}) \leqslant (v + 1)/M.$$

Therefore

$$\underline{|F(y_v+) - F(x)|}$$

$$= F(x) - F(y_v+) \leqslant F(y_{v+1}) - F(y_v+) \leqslant (v + 1)/M - v/M = \underline{1/M}$$

If event E occurs, then $|F_n(y_v+) - F(y_v+)| < c_1/8$. For the remaining inequality, we have

$$|F_n(x) - F_n(y_v+)| = F_n(x) - F_n(y_v+) \leqslant F_n(y_{v+1}) - F_n(y_v+)$$

$$\leqslant |F_n(y_{v+1}) - F(y_{v+1})| + |F(y_{v+1}) - F(y_v+)|$$

$$+ |F(y_v+) - F_n(y_v+)| < c_1/8 + 1/M + c_1/8,$$

assuming E has occurred.

In summary, $|F_n(x) - F(x)| < c_1/4 + 1/M + c_1/8 + 1/M$. Recall that $1/M < c_1/8$. Thus for $n \geqslant N$,

$$|F_n(x) - F(x)| < c_1$$

for all x, whenever E occurs. Event E^c occurs with probability less than c_2. It is therefore established that the number

$$N = N(c_1/8, c_2/(2(M + 1))),$$

where $1/M < c_1/8$, and the function N is as defined in the preceding lemma, suffices for the number N described in the theorem.

Appendix I

ANALYSIS, SOLUTION, AND COMPUTER PROGRAM FOR A SUPERVISED LEARNING PROBLEM IN PATTERN RECOGNITION

EXAMPLE.   The life of the PRM is 6 decision times, any number of which may be devoted to supervised learning, the remaining time being used for recognition. Find a learning strategy that minimizes the total Bayes risk of a machine which is defined by the following parameters: $N = 6$, $W = \{A, B\}$; $p(w) = \frac{1}{2}$, $w = A, B$.

For Pattern A,

$$p(z \mid \theta_A, A) = \begin{cases} \frac{1}{4} \text{ for } z = 0, \frac{3}{4} \text{ for } z = 1, & \text{if } \theta_A = 1, \\ \frac{3}{4} \text{ for } z = 0, \frac{1}{4} \text{ for } z = 1, & \text{if } \theta_A = 2. \end{cases}$$

For Pattern B,

$$p(z \mid \theta_B, B) = \begin{cases} \frac{1}{4} \text{ for } z = 0, \frac{3}{4} \text{ for } z = 1, & \text{if } \theta_B = 1, \\ \frac{3}{4} \text{ for } z = 0, \frac{1}{4} \text{ for } z = 1, & \text{if } \theta_B = 2. \end{cases}$$

$P[\theta_A = 1] = p[\theta_A = 2] = \frac{1}{2} = p[\theta_B = 1] = p[\theta_B = 2]$.
The recognition loss matrix is

		Actual pattern	
H(w, g):	g/w	A	B
Guess	A	$\begin{bmatrix} -1 \\ 1.5 \end{bmatrix}$	$\begin{bmatrix} 1.5 \\ -1 \end{bmatrix}$
	B		

(35)

ANALYSIS.   The pair (m_A, n_A) denotes that m_A 1's have occurred in n_A learning observations on Pattern A. (m_A, n_A) is therefore a sufficient statistic for θ_A [1, p. 389], and consequently the 4-tuple (m_A, n_A, m_B, n_B) is a sufficient statistic for $\theta = (\theta_A, \theta_B)$, and will serve as our state variable x. Our first project will be to reveal the equations for calculating $B(F(\theta \mid x))$.

From Bayes rule,

$$P(\theta_w \mid x) = p(x \mid \theta_w)p(\theta_w) \Big/ \left(\sum_{r \in \{1,2\}} p(x \mid r)p(r) \right), \qquad w = A, B; \theta_w = 1, 2.$$

By the "total probability" rule

$$p(z \mid w, x) = \sum_{\theta_w \in \{1,2\}} p(z \mid w, \theta_w)p(\theta_w \mid x).$$

Again by Bayes rule,

$$p(w \mid z, x) = p(z \mid w, x)p(w) \Big/ \left(\sum_{w' \in W} p(z \mid w', x)p(w') \right).$$

Define
$$K(w', z, x) = \sum_{w \in W} H(w', w)p(w \mid z, x)$$

and $d^*(z) = w^*$ such that $K(w^*, z, x) = \min_{w \in W} K(w, z, x)$.

$$p(z \mid x) = \sum_{w \in W} \sum_{\theta_w \in \{1,2\}} p(z \mid \theta_w, w)p(\theta_w \mid x)p(w),$$

and thus, according to the analysis of Chapter 5,

$$B\big(F(\theta \mid x)\big) = \sum_{z \in \{0,1\}} K(d^*(z), z, x)p(z \mid x)$$

has been defined in terms of the parameters of our particular process. Now we are in a position to proceed with the dynamic programming construction of a solution. At each decision time, the control set is

$$p = \{\text{Learn A, Learn B, Use}\}$$

or more simply, $\{A, B, U\}$. The loss function is determined by the equation

$$L(x, y, p, t) = \begin{cases} (N - t + 1)B\big(F(\theta \mid x)\big), & \text{if } p = \text{Use}, \\ 0, & \text{otherwise}. \end{cases}$$

Now the dynamic programming algorithm for adaptive control processes may be applied routinely. For $t = N = 6$, $S(x, 6) = p^*$, where $p^* = $ Use if $B\big(F(\theta \mid x)\big) < 0$, and p^* "Learn A or B" otherwise. The computer program to follow includes the control "give up" when it is impossible to reach a state for which $B\big(F(\theta \mid x)\big) < 0$. For $t < 6$, $S(x, t) = p^*$, where

$$H(x, p^*, t) = \min_{p \in P} H(x, p, t),$$

$H(x, p, t)$ being defined in the fashion of the dynamic programming algorithm. That is, suppose $x = (m_A, n_A, m_B, n_B)$. Then

$$H(x, A, t) = E[L((m_A + 1, n_A + 1, m_B, n_B), S, t + 1)]P(1 \mid A, x)$$
$$+ E[L((m_A, n_A + 1, m_B, n_B), S, t + 1)]P(0 \mid A, x),$$

where
$$P(1 \mid A, x) = \sum_{\theta_A \in \{1,2\}} P(1, \theta_A \mid x), \text{ etc.}$$

$H(x, B, t)$ is calculated similarly, and

$$H(x, U, t) = (N - t + 1)B\big(F(\theta \mid x)\big).$$

The computer program that follows performs these computations. In the printout, E[LOSS] tabulates $E[L(x, S, t)]$.

REFERENCE

1. Wilks, S. S. (1962). *Mathematical Statistics*. Wiley, New York.

```
*       COMPILE FORTRAN,EXECUTE FORTRAN
SUBROUTINEBINOM
 LAST DEFINED BUT NOT USED IN AN ARITH STMNT.
    N DEFINED BUT NOT USED IN AN ARITH STMNT.
*** MAIN PROGRAM ***
LEARN DEFINED BUT NOT USED IN AN ARITH STMNT.
  USE DEFINED BUT NOT USED IN AN ARITH STMNT.
   FN DEFINED BUT NOT USED IN AN ARITH STMNT.
   FM DEFINED BUT NOT USED IN AN ARITH STMNT.
   NW DEFINED BUT NOT USED IN AN ARITH STMNT.
   NK DEFINED BUT NOT USED IN AN ARITH STMNT.
  LNS DEFINED BUT NOT USED IN AN ARITH STMNT.
2000        SUBROUTINE BINOM

SUBROUTINE BINOM COMPUTES P(X/D) AND P(D/X).
            DIMENSION PW(2),H(2,2),P(6),PZWX(2,2),PZX(2),PWZX(2;2),TWZX(2,2),
           1TMIN(2),EL(3),PAT(2),ND(2),PD(2,6),PZWD(4,6),
           2PDX(2,6),DSTAR(2,2),ELOSS(1600,2),LABEL(1600,2)
            COMMON PW,H,P,PZWX,PZX,PWZX,TWZX,TMIN,ELOSS,EL,PAT,ND,PD,
           1PZWD,PDX,DSTAR,FN,FM,NW,LABEL
2005        FACTM=1.0
2007        FACTN=1.0
2009        LAST=FM-1.0
2013        IF(FM)40,10,40
2016     10 IF(FN)20,30,20
2018     20 FACTN=FN
2020     30 GO TO 60
2021     40 DO 50 NM=0,LAST
2025        FNM=NM
2028        FACTN=FACTN*(FN-FNM)
2032        FACTM=FACTM*(FM-FNM)
2036     50 CONTINUE
2037     60 BINC=FACTN/FACTM
2041        N=ND(NW)
2044        K=NW*2
2047        SUM=0.
2049        DO 70 I=1,N
2053        P(I)=BINC*(PZWD(K,I)**FM)*((1.0-PZWD(K,I))**(FN-FM))
2071        SUM=SUM+P(I)*PD(NW,I)
```

```
2079        70 CONTINUE
2080           DO 80 I=1,N
2084           PDX(NW,I)=P(I)*PD(NW,I)/SUM
2093        80 CONTINUE
2094           RETURN
               END
               DIMENSION PW(2),H(2,2),P(6),PZWX(2,2),PZX(2),PWZX(2,2),TWZX(2,2),
              1TMIN(2),EL(3),PAT(2),ND(2),PD(2,6),PZWD(4,6),
              2PDX(2,6),DSTAR(2,2),ELOSS(1600,2),LABEL(1600,2)
               COMMON PW,H,P,PZWX,PZX,PWZX,TWZX,TMIN,ELOSS,EL,PAT,ND,PD,
              1PZWD,PDX,DSTAR,FN,FM,NW,LABEL
2123         1 FORMAT(I2,A1,7X2A5,2A1,2I2,4X2F10.5)
2130         2 FORMAT(10X7F10.5)
2133         3 FORMAT(1X1H(I2,1H,I2,1H,I2,1H,I2,1H)5(2XF7.4),3(2XF10.6),2X
              17HGIVE UP)
2149         4 FORMAT(1X1H(I2,1H,I2,1H,I2,1H,I2,1H)5(2XF7.4),3(2XF10.6),2XA5,
              12(1XA1))
2163         5 FORMAT(1X1H(I2,1H,I2,1H,I2,1H,I2,1H)5(2XF7.4),3(2XF10.6),2XA5,5X
              12(1XA1),2X2(1XA1))
2180         6 FORMAT(1H13HK= I2//5X5HSTATE7X6HP(O/X)2X8HP(B/O,X)1X8HP(B/1,X)1X
              18HP(O/A,X)1X8HP(O/B,X)2X9HK(W',O,X)3X9HK(W',1,X)4X7HE(LOSS)4X
              26HS(X,1)3X5HD*(O)1X5HD*(1))
2212         9 FORMAT(15X10HP(D(A1)/X)8X10HP(D(B1)/X)//)
2221        11 FORMAT(1H139H*** PRM SUPERVISED LEARNING PROBLEM ***/2X27H** PROBL
              1EM SPECIFICATION **/5X23HNUMBER OF PATTERNS = 2 /5X14HMACHINE LIFE
              2 =I4,13H TIME PERIODS)
2251        12 FORMAT(/3X22H* PATTERN DISTRIBUTION/6X10HPATTERN(W)5X4HP(W)/
              12(/10XA1,9XF6.4))
2268        13 FORMAT(/3X13H* LOSS MATRIX/20X14HACTUAL PATTERN/23X1HA8X1HB//16X
              11HA2(3XF6.2)/8X5HGUESS/16X1HB2(3XF6.2))
2289        14 FORMAT(/3X26H* Z DISTRIBUTIONS,PATTERN A1/10X1HD10X4HP(D)6X
              16HP(O/D)5X6HP(1/D)/)
2307        15 FORMAT(10XI1,9XF6.4,2(5XF6.4))
2312        19 FORMAT(16XF7.4,11XF7.4/)
2316           EPSLN=.00005
2319       100 READ 1,390,NTIME,BLANK,LEARN,USE,A,B,ND(1),ND(2),PW(1),PW(2)
2338           PRINT 11,NTIME
2343           PRINT 12,A,PW(1),B,PW(2)
2354           DO 10 I=1,2
2357           NJ=ND(I)
2359        10 READ 2,(PD(I,J),J=1,NJ)
2374           DO 20 I=1,2
2377           NJ=ND(1)
2379        20 READ 2,(PZWD(I,J),J=1,NJ)
2394           PRINT 14,A
2399           DO 25 I=1,NJ
2403        25 PRINT 15,I,PD(1,I),PZWD(1,I),PZWD(2,I)
2421           DO 30 I=3,4
2424           NJ=ND(2)
2426        30 READ 2,(PZWD(I,J),J=1,NJ)
2441           PRINT 14,B
2446           DO 35 I=1,NJ
2450        35 PRINT 15,I,PD(2,I),PZWD(3,I),PZWD(4,I)
2468           DO 40 I=1,2
2471        40 READ 2,(H(I,J),J=1,2)
2485           PRINT 13,H(1,1),H(1,2),H(2,1),H(2,2)


2496           NC=1

    'DO 380' CONTROLS THE TIME PERIOD.
2498           DO 380 IN=1,NTIME
2502           NS=0
2504           NLINE=0
2506           NN=NTIME-IN
2509           KK=NN+1
2512           PRINT 6,KK
2517           PRINT 9
```

133

'DO 371' (NEST OF THREE) GENERATES THE STATE SET (M1,N1,M2,N2).

```
COMPUTE P(D/X).
2520          DO 371 INN=0,NN
2524          N1=NN-INN
2527          N2=KK-(N1+1)
2533          DO 371 M2=0,N2
2537          FN=N2
2540          FM=M2
2543          NW=2
2545          CALL BINOM
2546          DO 371 M1=0,N1
2550          FN=N1
2553          FM=M1
2556          NW=1
2558          CALL BINOM

COMPUTE P(Z/W,X).
2559          IJ=0
2561          DO 120 I=1,2
2564          DO 120 J=1,2
2567          IJ=IJ+1
2570          SUM=0.
2572          NK=ND(I)
2574          DO 110 K=1,NK
2578          SUM=SUM+PZWD(IJ,K)*PDX(I,K)
2590     110 CONTINUE
2591          PZWX(I,J)=SUM
2597     120 CONTINUE

COMPUTE P(Z/X).
2599          DO 140 J=1,2
2602          SUM=0.
2604          DO 130 I=1,2
2607          SUM=SUM+PZWX(I,J)*PW(I)
2615     130 CONTINUE
2616          PZX(J)=SUM
2618     140 CONTINUE

COMPUTE P(W/Z,X).
2619          DO 150 I=1,2
2622          DO 150 J=1,2
2625          PWZX(I,J)=PZWX(I,J)*PW(I)/PZX(J)
2634     150 CONTINUE
```

```
      COMPUTE K(W,Z,X).
2636          DO 170 J=1,2
2639          DO 170 I=1,2
2642          SUM=0.
2644          DO 160 K=1,2
2647          SUM=SUM+H(K,I)*PWZX(K,J)
2659      160 CONTINUE
2660          TWZX(I,J)=SUM
2666      170 CONTINUE

      DETERMINE K(W',Z,X),W'.
2668          DO 190 J=1,2
2671          TMIN(J)=TWZX(1,J)
2676          DO 190 I=1,2
2679          IF(TMIN(J)-TWZX(I,J))190,190,180
2687      180 TMIN(J)=TWZX(I,J)
2689      190 CONTINUE

      DETERMINE D*(Z).
2691          DO 195 I=1,4
2694          DSTAR(I)=BLANK
2696      195 CONTINUE
2697          DO 210 J=1,2
2700          NJ=0
2702          DO 210 I=1,2
2705          IF(TMIN(J)-TWZX(I,J))210,200,210
2713      200 NJ=NJ+1
2716          GO TO(205,206),I
2721      205 DSTAR(J,NJ)=A
2727          GO TO 210
2728      206 DSTAR(J,NJ)=B
2734      210 CONTINUE
2736          NLINE=NLINE+3
2739          IF(NLINE-57)219,218,218
2743      218 PRINT 6,KK
2748          PRINT 9
2751          NLINE=3
2753      219 NS=NS+1
2756          LABEL(NS,NC)=N2+M2*100+N1*10000+M1*1000000
2772          IF(IN-1)280,220,280
```

```
      COMPUTE EXPECTED LOSS IN FINAL DECISION PERIOD.
2776      220 SUM=0.
2778          DO 230 J=1,2
2781          SUM=SUM+TMIN(J)*PZX(J)
2785      230 CONTINUE
2786          IF(SUM-EPSLN)240,260,260
2791      240 IF(-EPSLN-SUM)260,260,250
2795      250 ELOSS(NS,NC)=SUM
2801          PRINT 5,M1,N1,M2,N2,PZX(1),PWZX(2,1),PWZX(2,2),PZWX(1,1),
             1PZWX(2,1),TMIN(1),TMIN(2),ELOSS(NS,NC),USE,(DSTAR(1,K),K=1,2),
             2(DSTAR(2,K),K=1,2)
2845          PRINT 19,PDX(1,1),PDX(2,1)
2852          GO TO 371
2853      260 ELOSS(NS,NC)=0.
2859      270 PRINT 3,M1,N1,M2,N2,PZX(1),PWZX(2,1),PWZX(2,2),PZWX(1,1),
             1PZWX(2,1),TMIN(1),TMIN(2),ELOSS(NS,NC)
2883          PRINT 19,PDX(1,1),PDX(2,1)
2890          GO TO 371

      COMPUTE EXPECTED LOSS IF 'USE'.
2891      280 EL(1)=0.
2893          DO 290 J=1,2
2896          EL(1)=EL(1)+TMIN(J)*PZX(J)
2900      290 CONTINUE
2901          RD=NTIME-KK+1
2906          EL(1)=RD*EL(1)

      COMPUTE EXPECTED LOSS IF 'LEARN A'.
2909          EL(2)=0.
2911          ICON=N2+M2*100+(N1+1)*10000
2922          DO 297 I=1,2
2925          LAST=ICON+(M1-1+I)*1000000
2932          DO 295 J=1,LNS
2936          IF(LAST-LABEL(J,LNC))295,296,295
2943      295 CONTINUE
2944      296 EL(2)=EL(2)+ELOSS(J,LNC)*PZWX(1,I)
2955      297 CONTINUE
```

```
       COMPUTE EXPECTED LOSS IF 'LEARN B'.
2956          EL(3)=0.
2958          ICON=(N2+1)+N1*10000+M1*1000000
2969          DO 300 I=1,2
2972          LAST=ICON+(M2-1+I)*100
2979          DO 298 J=1,LNS
2983          IF(LAST-LABEL(J,LNC))298,299,298
2990      298 CONTINUE
2991      299 EL(3)=EL(3)+ELOSS(J,LNC)*PZWX(2,I)
3002      300 CONTINUE

       DETERMINE DECISION,S(X,T).
3003          IJ=0
3005          DO 305 I=1,3
3008          IF(EL(I)-EPSLN)301,303,303
3013      301 IF(-EPSLN-EL(I))302,305,305
3018      302 EL(I)=0.
3020      303 IJ=IJ+1
3023      305 CONTINUE
3024          IF(IJ-3)307,306,307
3028      306 ELOSS(NS,NC)=0.
3034          PRINT 3,M1,N1,M2,N2,PZX(1),PWZX(2,1),PWZX(2,2),PZWX(1,1),
              1PZWX(2,1),TMIN(1),TMIN(2),ELOSS(NS,NC)
3058          PRINT 19,PDX(1,1),PDX(2,1)
3065          GO TO 371
3066      307 ELMIN=EL(1)
3068          DO 320 K=2,3
3071          IF(ELMIN-EL(K))320,320,310
3075      310 ELMIN=EL(K)
3077      320 CONTINUE
3078          ELOSS(NS,NC)=ELMIN
3084          IJ=0
3086          DO 370 K=1,3
3089          IF(ELMIN-EL(K))370,330,370
3093      330 GO TO(340,350,360),K
3099      340 PRINT 5,M1,N1,M2,N2,PZX(1),PWZX(2,1),PWZX(2,2),PZWX(1,1),
              1PZWX(2,1),TMIN(1),TMIN(2),ELOSS(NS,NC),USE,(DSTAR(1,K),K=1,2),
              2(DSTAR(2,K),K=1,2)
3147          PRINT 19,PDX(1,1),PDX(2,1)
3154          GO TO 371
3155      350 IJ=IJ+1
3158          PAT(IJ)=A
3161          GO TO 370
3162      360 IJ=IJ+1
3165          PAT(IJ)=B
```

```
3168    370 CONTINUE
3169        PRINT 4,M1,N1,M2,N2,PZX(1),PWZX(2,1),PWZX(2,2),PZWX(1,1),
            1PZWX(2,1),TMIN(1),TMIN(2),ELOSS(NS,NC),LEARN,(PAT(I),I=1,IJ)
3206        PRINT 19,PDX(1,1),PDX(2,1)
3213    371 CONTINUE
3216        LNS=NS
3218        GO TO (375,376),NC
3223    375 NC=2
3225        LNC=1
3227        GO TO 380
3228    376 NC=1
3230        LNC=2
3232    380 CONTINUE
3233        GO TO 100
3234    390 STOP
            END
```

```
            *** PRM SUPERVISED LEARNING PROBLEM ***
            ** PROBLEM SPECIFICATION **
            NUMBER OF PATTERNS = 2
            MACHINE LIFE =   6 TIME PERIODS

            * PATTERN DISTRIBUTION
            PATTERN(W)     P(W)

                A          0.5000
                B          0.5000

            * Z DISTRIBUTIONS,PATTERN A
                D          P(D)      P(0/D)      P(1/D)

                1          0.5000    0.2500      0.7500
                2          0.5000    0.7500      0.2500

            * Z DISTRIBUTIONS,PATTERN B
                D          P(D)      P(0/D)      P(1/D)

                1          0.5000    0.2500      0.7500
                2          0.5000    0.7500      0.2500

            * LOSS MATRIX
                           ACTUAL PATTERN
                           A         B

                    A     -1.00      1.50
              GUESS
                    B      1.50     -1.00
```

STATE	P(O/X) / P(C(A1)/X)	P(B/O,X)	P(B/1,X) / P(D(B1)/X)	P(O/A,X)	P(O/B,X)	K(W',O,X)	K(W',1,X)	E(LOSS)	S(X,1)	D*(O)	D*(1)
(0, 5, 0, 0)	0.6240 / 0.0041	0.4007	0.6649 / 0.5000	0.7480	0.5000	0.001642	-0.162125	-0.059939	USE	A	B
(1, 5, 0, 0)	0.6161 / 0.0357	0.4058	0.6512 / 0.5000	0.7321	0.5000	0.014493	-0.127907	-0.040179	USE	A	B
(2, 5, 0, 0)	0.5625 / 0.2500	0.4444	0.5714 / 0.5000	0.6250	0.5000	0.111111	0.071429	0.000000	GIVE UP		
(3, 5, 0, 0)	0.4375 / 0.7500	0.5714	0.4444 / 0.5000	0.3750	0.5000	0.071429	0.111111	0.000000	GIVE UP		
(4, 5, 0, 0)	0.3839 / 0.9643	0.6512	0.4058 / 0.5000	0.2679	0.5000	-0.127907	0.014493	-0.040179	USE	B	A
(5, 5, 0, 0)	0.3760 / 0.9959	0.6649	0.4007 / 0.5000	0.2520	0.5000	-0.162125	0.001642	-0.059939	USE	B	A
(0, 4, 0, 1)	0.6845 / 0.0122	0.4566	0.5942 / 0.2500	0.7439	0.6250	0.141425	0.014493	0.000000	GIVE UP		
(1, 4, 0, 1)	0.6625 / 0.1000	0.4717	0.5556 / 0.2500	0.7000	0.6250	0.179245	0.111111	0.000000	GIVE UP		
(2, 4, 0, 1)	0.5625 / 0.5000	0.5556	0.4286 / 0.2500	0.5000	0.6250	0.111111	0.071428	0.000000	GIVE UP		
(3, 4, 0, 1)	0.4625 / 0.9000	0.6757	0.3488 / 0.2500	0.3000	0.6250	-0.189189	-0.127907	-0.156250	USE	B	A
(4, 4, 0, 1)	0.4405 / 0.9878	0.7093	0.3351 / 0.2500	0.2561	0.6250	-0.273356	-0.162125	-0.211128	USE	B	A
(0, 4, 1, 1)	0.5595 / 0.0122	0.3351	0.7093 / 0.7500	0.7439	0.3750	-0.162125	-0.273356	-0.211128	USE	A	B
(1, 4, 1, 1)	0.5375 / 0.1000	0.3488	0.6757 / 0.7500	0.7000	0.3750	-0.127907	-0.189189	-0.156250	USE	A	B

K= 6

STATE	P(O/X) P(D(A1)/X)	P(B/O,X)	P(B/1,X) P(D(B1)/X)	P(O/A,X)	P(O/B,X)	K(W',0,X)	K(W',1,X)	E(LOSS)	S(X,1)	D*(0)	D*(1)
(2, 4, 1, 1)	0.4375 0.5000	0.4286	0.5556 0.7500	0.5000	0.3750	0.071428	0.111111	0.000000	GIVE UP		
(3, 4, 1, 1)	0.3375 0.9000	0.5556	0.4717 0.7500	0.3000	0.3750	0.111111	0.179245	0.000000	GIVE UP		
(4, 4, 1, 1)	0.3155 0.9878	0.5942	0.4566 0.7500	0.2561	0.3750	0.014493	0.141425	0.000000	GIVE UP		
(0, 3, 0, 2)	0.7161 0.0357	0.4888	0.5283 0.1000	0.7321	0.7000	0.221945	0.179245	0.000000	GIVE UP		
(1, 3, 0, 2)	0.6625 0.2500	0.5283	0.4444 0.1000	0.6250	0.7000	0.179245	0.111111	0.000000	GIVE UP		
(2, 3, 0, 2)	0.5375 0.7500	0.6512	0.3243 0.1000	0.3750	0.7000	-0.127907	-0.189189	-0.156250	USE	B	A
(3, 3, 0, 2)	0.4839 0.9643	0.7232	0.2907 0.1000	0.2679	0.7000	-0.308818	-0.273356	-0.290179	USE	B	A
(0, 3, 1, 2)	0.6161 0.0357	0.4058	0.6512 0.5000	0.7321	0.5000	0.014493	-0.127907	-0.040179	USE	A	B
(1, 3, 1, 2)	0.5625 0.2500	0.4444	0.5714 0.5000	0.6250	0.5000	0.111111	0.071429	0.000000	GIVE UP		
(2, 3, 1, 2)	0.4375 0.7500	0.5714	0.4444 0.5000	0.3750	0.5000	0.071429	0.111111	0.000000	GIVE UP		
(3, 3, 1, 2)	0.3839 0.9643	0.6512	0.4058 0.5000	0.2679	0.5000	-0.127907	0.014493	-0.040179	USE	B	A

K= 6

STATE	P(O/X) P(D(A1)/X)	P(B/O,X)	P(B/1,X) P(D(B1)/X)	P(O/A,X)	P(O/B,X)	K(W',O,X)	K(W',1,X)	E(LOSS)	S(X,1)	D*(0)	D*(1)
(0, 3, 2, 2)	0.5161 0.0357	0.2907	0.7232 0.9000	0.7321	0.3000	-0.273356	-0.308118	-0.290179	USE	A	B
(1, 3, 2, 2)	0.4625 0.2500	0.3243	0.6512 0.9000	0.6250	0.3000	-0.189189	-0.127907	-0.156250	USE	A	B
(2, 3, 2, 2)	0.3375 0.7500	0.4444	0.5283 0.9000	0.3750	0.3000	0.111111	0.179245	0.000000	GIVE UP		
(3, 3, 2, 2)	0.2839 0.9643	0.5283	0.4888 0.9000	0.2679	0.3000	0.179245	0.221945	0.000000	GIVE UP		
(0, 2, 0, 3)	0.7161 0.1000	0.5112	0.4717 0.0357	0.7000	0.7321	0.221945	0.179245	0.000000	GIVE UP		
(1, 2, 0, 3)	0.6161 0.5000	0.5942	0.3488 0.0357	0.5000	0.7321	0.014493	-0.127907	-0.040179	USE	B	A
(2, 2, 0, 3)	0.5161 0.9000	0.7093	0.2768 0.0357	0.3000	0.7321	-0.273356	-0.308118	-0.290179	USE	B	A
(0, 2, 1, 3)	0.6625 0.1000	0.4717	0.5556 0.2500	0.7000	0.6250	0.179245	0.111111	0.000000	GIVE UP		
(1, 2, 1, 3)	0.5625 0.5000	0.5556	0.4286 0.2500	0.5000	0.6250	0.111111	0.071429	0.000000	GIVE UP		
(2, 2, 1, 3)	0.4625 0.9000	0.6757	0.3488 0.2500	0.3000	0.6250	-0.189189	-0.127907	-0.156250	USE	B	A
(0, 2, 2, 3)	0.5375 0.1000	0.3488	0.6757 0.7500	0.7000	0.3750	-0.127907	-0.189189	-0.156250	USE	A	B
(1, 2, 2, 3)	0.4375 0.5000	0.4286	0.5556 0.7500	0.5000	0.3750	0.071429	0.111111	0.000000	GIVE UP		

K= 6

STATE	P(O/X) / P(D(A1)/X)	P(B/O,X)	P(B/1,X) / P(D(B1)/X)	P(O/A,X)	P(O/B,X)	K(W',0,X)	K(W',1,X)	E(LOSS)	S(X,1)	D*(0)	D*(1)
(2, 2, 2, 3)	0.3375 / 0.9000	0.5556	0.4717 / 0.7500	0.3000	0.3750	0.111111	0.179245	0.000000	GIVE UP		
(0, 2, 3, 3)	0.4839 / 0.1000	0.2768	0.7093 / 0.9643	0.7000	0.2679	-0.308118	-0.273356	-0.290179	USE	A	B
(1, 2, 3, 3)	0.3839 / 0.5000	0.3488	0.5942 / 0.9643	0.5000	0.2679	-0.127907	0.014493	-0.040179	USE	A	B
(2, 2, 3, 3)	0.2839 / 0.9000	0.4717	0.5112 / 0.9643	0.3000	0.2679	0.179245	0.221945	0.000000	GIVE UP		
(0, 1, 0, 4)	0.6845 / 0.2500	0.5434	0.4058 / 0.0122	0.6250	0.7439	0.141425	0.014493	0.000000	GIVE UP		
(1, 1, 0, 4)	0.5595 / 0.7500	0.6649	0.2907 / 0.0122	0.3750	0.7439	-0.162125	-0.273356	-0.211128	USE	B	A
(0, 1, 1, 4)	0.6625 / 0.2500	0.5283	0.4444 / 0.1000	0.6250	0.7000	0.179245	0.111111	0.000000	GIVE UP		
(1, 1, 1, 4)	0.5375 / 0.7500	0.6512	0.3243 / 0.1000	0.3750	0.7000	-0.127907	-0.189189	-0.156250	USE	B	A
(0, 1, 2, 4)	0.5625 / 0.2500	0.4444	0.5714 / 0.5000	0.6250	0.5000	0.111111	0.071428	0.000000	GIVE UP		
(1, 1, 2, 4)	0.4375 / 0.7500	0.5714	0.4444 / 0.5000	0.3750	0.5000	0.071428	0.111111	0.000000	GIVE UP		
(0, 1, 3, 4)	0.4625 / 0.2500	0.3243	0.6512 / 0.9000	0.6250	0.3000	-0.189189	-0.127907	-0.156250	USE	A	B
(1, 1, 3, 4)	0.3375 / 0.7500	0.4444	0.5283 / 0.9000	0.3750	0.3000	0.111111	0.179245	0.000000	GIVE UP		

K= 6

STATE	P(O/X) / P(C(A1)/X)	P(B/O,X)	P(B/1,X) / P(D(B1)/X)	P(O/A,X)	P(O/B,X)	K(W',0,X)	K(W',1,X)	E(LOSS)	S(X,1)	D*(0)	D*(1)
(0, 1, 4, 4)	0.4405 / 0.2500	0.2907	0.6649 / 0.9878	0.6250	0.2561	-0.273356	-0.162125	-0.211128	USE	A	B
(1, 1, 4, 4)	0.3155 / 0.7500	0.4058	0.5434 / 0.9878	0.3750	0.2561	0.014493	0.141425	0.000000	GIVE UP		
(0, 0, 0, 5)	0.6240 / 0.5000	0.5993	0.3351 / 0.0041	0.5000	0.7480	0.001642	-0.162125	-0.059939	USE	B	A
(0, 0, 1, 5)	0.6161 / 0.5000	0.5942	0.3488 / 0.0357	0.5000	0.7321	0.014493	-0.127907	-0.040179	USE	B	A
(0, 0, 2, 5)	0.5625 / 0.5000	0.5556	0.4286 / 0.2500	0.5000	0.6250	0.111111	0.071429	0.000000	GIVE UP		
(0, 0, 3, 5)	0.4375 / 0.5000	0.4286	0.5556 / 0.7500	0.5000	0.3750	0.071429	0.111111	0.000000	GIVE UP		
(0, 0, 4, 5)	0.3839 / 0.5000	0.3488	0.5942 / 0.9643	0.5000	0.2679	-0.127907	0.014493	-0.040179	USE	A	B
(0, 0, 5, 5)	0.3760 / 0.5000	0.3351	0.5993 / 0.9959	0.5000	0.2520	-0.162125	0.001642	-0.059939	USE	A	B

143

K= 5

STATE	P(O/X) / P(D(A1)/X)	P(B/O,X)	P(B/1,X) / P(D(B1)/X)	P(O/A,X)	P(O/B,X)	K(W',O,X)	K(W',1,X)	E(LOSS)	S(X,1)	D*(0)	D*(1)
(0, 4, 0, 0)	0.6220 / 0.0122	0.4020	0.6613 / 0.5000	0.7439	0.5000	0.004902	-0.153226	-0.109756	USE	A	B
(1, 4, 0, 0)	0.6000 / 0.1000	0.4167	0.6250 / 0.5000	0.7000	0.5000	0.041667	-0.062500	-0.078125	LEARN B		
(2, 4, 0, 0)	0.5000 / 0.5000	0.5000	0.5000 / 0.5000	0.5000	0.5000	0.250000	0.250000	0.000000	GIVE UP		
(3, 4, 0, 0)	0.4000 / 0.9000	0.6250	0.4167 / 0.5000	0.3000	0.5000	-0.062500	0.041667	-0.078125	LEARN B		
(4, 4, 0, 0)	0.3780 / 0.9878	0.6613	0.4020 / 0.5000	0.2561	0.5000	-0.153226	0.004902	-0.109756	USE	B	A
(0, 3, 0, 1)	0.6786 / 0.0357	0.4605	0.5833 / 0.2500	0.7321	0.6250	0.151316	0.041667	-0.015067	LEARN B		
(1, 3, 0, 1)	0.6250 / 0.2500	0.5000	0.5000 / 0.2500	0.6250	0.6250	0.250000	0.250000	0.000000	GIVE UP		
(2, 3, 0, 1)	0.5000 / 0.7500	0.6250	0.3750 / 0.2500	0.3750	0.6250	-0.062500	-0.062500	-0.125000	USE	B	A
(3, 3, 0, 1)	0.4464 / 0.9643	0.7000	0.3387 / 0.2500	0.2679	0.6250	-0.250000	-0.153226	-0.392857	USE	B	A
(0, 3, 1, 1)	0.5536 / 0.0357	0.3387	0.7000 / 0.7500	0.7321	0.3750	-0.153226	-0.250000	-0.392857	USE	A	B
(1, 3, 1, 1)	0.5000 / 0.2500	0.3750	0.6250 / 0.7500	0.6250	0.3750	-0.062500	-0.062500	-0.125000	USE	A	B
(2, 3, 1, 1)	0.3750 / 0.7500	0.5000	0.5000 / 0.7500	0.3750	0.3750	0.250000	0.250000	0.000000	GIVE UP		

STATE	P(O/X) P(C(A1)/X)	P(B/O,X)	P(B/1,X) P(D(B1)/X)	P(O/A,X)	P(O/B,X)	K(W',O,X)	K(W',1,X)	E(LOSS)	S(X,1)	D*(0)	D*(1)
(3, 3, 1, 1)	0.3214 0.9643	0.5833	0.4605 0.7500	0.2679	0.3750	0.041667	0.151316	-0.015067	LEARN B		
(0, 2, 0, 2)	0.7000 0.1000	0.5000	0.5000 0.1000	0.7000	0.7000	0.250000	0.250000	0.000000	GIVE UP		
(1, 2, 0, 2)	0.6000 0.5000	0.5833	0.3750 0.1000	0.5000	0.7000	0.041667	-0.062500	-0.078125	LEARN A		
(2, 2, 0, 2)	0.5000 0.9000	0.7000	0.3000 0.1000	0.3000	0.7000	-0.250000	-0.250000	-0.500000	USE	B	A
(0, 2, 1, 2)	0.6000 0.1000	0.4167	0.6250 0.5000	0.7000	0.5000	0.041667	-0.062500	-0.078125	LEARN B		
(1, 2, 1, 2)	0.5000 0.5000	0.5000	0.5000 0.5000	0.5000	0.5000	0.250000	0.250000	0.000000	GIVE UP		
(2, 2, 1, 2)	0.4000 0.9000	0.6250	0.4167 0.5000	0.3000	0.5000	-0.062500	0.041667	-0.078125	LEARN B		
(0, 2, 2, 2)	0.5000 0.1000	0.3000	0.7000 0.9000	0.7000	0.3000	-0.250000	-0.250000	-0.500000	USE	A	B
(1, 2, 2, 2)	0.4000 0.5000	0.3750	0.5833 0.9000	0.5000	0.3000	-0.062500	0.041667	-0.078125	LEARN A		
(2, 2, 2, 2)	0.3000 0.9000	0.5000	0.5000 0.9000	0.3000	0.3000	0.250000	0.250000	0.000000	GIVE UP		
(0, 1, 0, 3)	0.6786 0.2500	0.5395	0.4167 0.0357	0.6250	0.7321	0.151316	0.041667	-0.015067	LEARN A		
(1, 1, 0, 3)	0.5536 0.7500	0.6613	0.3000 0.0357	0.3750	C.7321	-0.153226	-0.250000	-0.392857	USE	B	A
(0, 1, 1, 3)	0.6250 0.2500	0.5000	0.5000 0.2500	0.6250	0.6250	0.250000	0.250000	0.000000	GIVE UP		

145

K= 5

STATE	P(O/X) / P(D(A1)/X)	P(B/O,X)	P(B/1,X) / P(D(B1)/X)	P(O/A,X)	P(O/B,X)	K(W',0,X)	K(W',1,X)	E(LOSS)	S(X,1)	D*(0)	D*(1)
(1, 1, 1, 3)	0.5000 / 0.7500	0.6250	0.3750 / 0.2500	0.3750	0.6250	-0.062500	-0.062500	-0.125000	USE	B	A
(0, 1, 2, 3)	0.5000 / 0.2500	0.3750	0.6250 / 0.7500	0.6250	0.3750	-0.062500	-0.062500	-0.125000	USE	A	B
(1, 1, 2, 3)	0.3750 / 0.7500	0.5000	0.5000 / 0.7500	0.3750	0.3750	0.250000	0.250000	0.000000	GIVE UP		
(0, 1, 3, 3)	0.4464 / 0.2500	0.3000	0.6613 / 0.9643	0.6250	0.2679	-0.250000	-0.153226	-0.392857	USE	A	B
(1, 1, 3, 3)	0.3214 / 0.7500	0.4167	0.5395 / 0.9643	0.3750	0.2679	0.041667	0.151316	-0.015067	LEARN A		
(0, 0, 0, 4)	0.6220 / 0.5000	0.5980	0.3387 / 0.0122	0.5000	0.7439	0.004902	-0.153226	-0.109756	USE	B	A
(0, 0, 1, 4)	0.6000 / 0.5000	0.5833	0.3750 / 0.1000	0.5000	0.7000	0.041667	-0.062500	-0.078125	LEARN A		
(0, 0, 2, 4)	0.5000 / 0.5000	0.5000	0.5000 / 0.5000	0.5000	0.5000	0.250000	0.250000	0.000000	GIVE UP		
(0, 0, 3, 4)	0.4000 / 0.5000	0.3750	0.5833 / 0.9000	0.5000	0.3000	-0.062500	0.041667	-0.078125	LEARN A		
(0, 0, 4, 4)	0.3780 / 0.5000	0.3387	0.5980 / 0.9878	0.5000	0.2561	-0.153226	0.004902	-0.109756	USE	A	B

K= 4

STATE	P(O/X) / P((A1)/X)	P(B/O,X)	P(B/1,X) / P(D(B1)/X)	P(O/A,X)	P(O/B,X)	K(W',0,X)	K(W',1,X)	E(LOSS)	S(X,1)	D*(0)	D*(1)
(0, 3, 0, 0)	0.6161 / 0.0357	0.4058	0.6512 / 0.5000	0.7321	0.5000	0.014493	-0.127907	-0.203962	LEARN B		
(1, 3, 0, 0)	0.5625 / 0.2500	0.4444	0.5714 / 0.5000	0.6250	0.5000	0.111111	0.071429	-0.062500	LEARN B		
(2, 3, 0, 0)	0.4375 / 0.7500	0.5714	0.4444 / 0.5000	0.3750	0.5000	0.071429	0.111111	-0.062500	LEARN B		
(3, 3, 0, 0)	0.3839 / 0.9643	0.6512	0.4058 / 0.5000	0.2679	0.5000	-0.127907	0.014493	-0.203962	LEARN B		
(0, 2, 0, 1)	0.6625 / 0.1000	0.4717	0.5556 / 0.2500	0.7000	0.6250	0.179245	0.111111	-0.029297	LEARN B		
(1, 2, 0, 1)	0.5625 / 0.5000	0.5556	0.4286 / 0.2500	0.5000	0.6250	0.111111	0.071429	-0.062500	LEARN A		
(2, 2, 0, 1)	0.4625 / 0.9000	0.6757	0.3488 / 0.2500	0.3000	0.6250	-0.189189	-0.127907	-0.468750	USE	B	A
(0, 2, 1, 1)	0.5375 / 0.1000	0.3488	0.6757 / 0.7500	0.7000	0.3750	-0.127907	-0.189189	-0.468750	USE	A	B
(1, 2, 1, 1)	0.4375 / 0.5000	0.4286	0.5556 / 0.7500	0.5000	0.3750	0.071429	0.111111	-0.062500	LEARN A		
(2, 2, 1, 1)	0.3375 / 0.9000	0.5556	0.4717 / 0.7500	0.3000	0.3750	0.111111	0.179245	-0.029297	LEARN B		
(0, 1, 0, 2)	0.6625 / 0.2500	0.5283	0.4444 / 0.1000	0.6250	0.7000	0.179245	0.111111	-0.029297	LEARN A		
(1, 1, 0, 2)	0.5375 / 0.7500	0.6512	0.3243 / 0.1000	0.3750	0.7000	-0.127907	-0.189189	-0.468750	USE	B	A

K= 4

STATE	P(O/X) P(C(A1)/X)	P(B/O,X)	P(B/1,X) P(D(B1)/X)	P(O/A,X)	P(O/B,X)	K(W',0,X)	K(W',1,X)	E(LOSS)	S(X,1)	D*(0)	D*(1)
(0, 1, 1, 2)	0.5625 / 0.2500	0.4444	0.5714 / 0.5000	0.6250	0.5000	0.111111	0.071429	-0.062500	LEARN B		
(1, 1, 1, 2)	0.4375 / 0.7500	0.5714	0.4444 / 0.5000	0.3750	0.5000	0.071429	0.111111	-0.062500	LEARN B		
(0, 1, 2, 2)	0.4625 / 0.2500	0.3243	0.6512 / 0.9000	0.6250	0.3000	-0.189189	-0.127907	-0.468750	USE	A	B
(1, 1, 2, 2)	0.3375 / 0.7500	0.4444	0.5283 / 0.9000	0.3750	0.3000	0.111111	0.179245	-0.029297	LEARN A		
(0, 0, 0, 3)	0.6161 / 0.5000	0.5942	0.3488 / 0.0357	0.5000	0.7321	0.014493	-0.127907	-0.203962	LEARN A		
(0, 0, 1, 3)	0.5625 / 0.5000	0.5556	0.4286 / 0.2500	0.5000	0.6250	0.111111	0.071429	-0.062500	LEARN A		
(0, 0, 2, 3)	0.4375 / 0.5000	0.4286	0.5556 / 0.7500	0.5000	0.3750	0.071429	0.111111	-0.062500	LEARN A		
(0, 0, 3, 3)	0.3839 / 0.5000	0.3488	0.5942 / 0.9643	0.5000	0.2679	-0.127907	0.014493	-0.203962	LEARN A		

K= 3

STATE	P(O/X) P(D(A1)/X)	P(B/O,X)	P(B/1,X) P(D(B1)/X)	P(O/A,X)	P(O/B,X)	K(W',O,X)	K(W',1,X)	E(LOSS)	S(X,1)	D*(0) D*(1)
(0, 2, 0, 0)	0.6000 0.1000	0.4167	0.6250 0.5000	0.7000	0.5000	0.041667	-0.062500	-0.249023	LEARN B	
(1, 2, 0, 0)	0.5000 0.5000	0.5000	0.5000 0.5000	0.5000	0.5000	0.250000	0.250000	-0.062500	LEARN A B	
(2, 2, 0, 0)	0.4000 0.9000	0.6250	0.4167 0.5000	0.3000	0.5000	-0.062500	0.041667	-0.249023	LEARN B	
(0, 1, 0, 1)	0.6250 0.2500	0.5000	0.5000 0.2500	0.6250	0.6250	0.250000	0.250000	-0.041748	LEARN A B	
(1, 1, 0, 1)	0.5000 0.7500	0.6250	0.3750 0.2500	0.3750	0.6250	-0.062500	-0.062500	-0.316406	LEARN B	
(0, 1, 1, 1)	0.5000 0.2500	0.3750	0.6250 0.7500	0.6250	0.3750	-0.062500	-0.062500	-0.316406	LEARN A	
(1, 1, 1, 1)	0.3750 0.7500	0.5000	0.5000 0.7500	0.3750	0.3750	0.250000	0.250000	-0.041748	LEARN A B	
(0, 0, 0, 2)	0.6000 0.5000	0.5833	0.3750 0.1000	0.5000	0.7000	0.041667	-0.062500	-0.249023	LEARN A	
(0, 0, 1, 2)	0.5000 0.5000	0.5000	0.5000 0.5000	0.5000	0.5000	0.250000	0.250000	-0.062500	LEARN A B	
(0, 0, 2, 2)	0.4000 0.5000	0.3750	0.5833 0.9000	0.5000	0.3000	-0.062500	0.041667	-0.249023	LEARN A	

149

K= 2

STATE	P(O/X) P(D(A1)/X)	P(B/0,X)	P(B/1,X) P(D(B1)/X)	P(O/A,X)	P(O/B,X)	K(W',0,X)	K(W',1,X)	E(LOSS)	S(X,1)	D*(0) D*(1)
(0, 1, 0, 0)	0.5625 0.2500	0.4444	0.5714 0.5000	0.6250	0.5000	0.111111	0.071428	-0.179077	LEARN A	
(1, 1, 0, 0)	0.4375 0.7500	0.5714	0.4444 0.5000	0.3750	0.5000	0.071428	0.111111	-0.179077	LEARN A B	
(0, 0, 0, 1)	0.5625 0.5000	0.5556	0.4286 0.2500	0.5000	0.6250	0.111111	0.071428	-0.179077	LEARN B	
(0, 0, 1, 1)	0.4375 0.5000	0.4286	0.5556 0.7500	0.5000	0.3750	0.071428	0.111111	-0.179077	LEARN A B	

K= 1

STATE	P(O/X) P(D(A1)/X)	P(B/0,X)	P(B/1,X) P(D(B1)/X)	P(O/A,X)	P(O/B,X)	K(W',0,X)	K(W',1,X)	E(LOSS)	S(X,1)	D*(0) D*(1)
(0, 0, 0, 0)	0.5000 0.5000	0.5000	0.5000 0.5000	0.5000	0.5000	0.250000	0.250000	-0.179077	LEARN A B	

LIST OF SYMBOLS

LIST OF SYMBOLS

Symbol	Pages where defined	Symbol represents
A	6	Set of decision times
B(F)	79	Bayes Risk with respect to F.
E[·]		The expectation operator
E[· | ·]		The conditional expectation operator
E[L(x, S, t)]	28	Expected loss using strategy S at initial state-time (x, t).
F(·)	125	Cumulative distribution function
F(· | ·)	80	Conditional cumulative distribution function
$F_n(·)$	74	Empirical distribution function
G(x̄, p̄, t)	50	Gain function
H(w, g)	72	Loss matrix for a PRM
H(x, p, t)	35	
I(W, V)	84	Channel transinformation
K(z, w*)	79	
L	77	Sequence of learning observations
L(x, p̄, t)	7	Loss function
L(x̄, p̄, t)	27	Generalized loss function
L(x, y, p, t)	30	
O(x, t)	35	
p	6	Control element
p̄	7	Policy
p*	14	
P(x, t)	6	Control set function
P(· | ·)		Conditional probability
Q(i)	8	Condition on state-space
r(m, n)	51	
S(x, t)	27	Strategy
S_m	35	
t	6	Time
T(x, p, t)	6	Law of motion
v̄(y)	14	
w	72	Pattern
w*	72	

LIST OF SYMBOLS (*continued*)

Symbol	Pages where defined	Symbol represents
W	72	The set of patterns
x	6	State
$\bar{x}$	7	Trajectory
X'_t	28	
Z	72	Transducer measurement
z	72	Observation of transducer measurement
θ_i	73	Parameter index function for a pattern-conditional distribution
θ	76	Parameter index function for all pattern-conditional distributions
Ω	6	State-space
$\mathscr{F}$	27	Statistical law of motion
$\mathscr{F}_i$	73	Set of pattern-conditional c.d.f.'s
$\mathscr{S}$	29	Set of strategies
= over a letter	15	Cardinal number of the set represented by the letter
$\times$		Cartesian product

AUTHOR INDEX

Page numbers set in *italics* denote the pages on which
the complete literature references are given.

SUBJECT INDEX

Markov property, 2, 30
Mathematical modeling, 1
Minimax criterion, 30
Modeling, 1
Modified problem, 13
  solution to, 13
Monotonicity property, 23

Operations research, 4
Order statistic, 74

Pattern recognition machine
  see also supervised learning
  model, 4, 72
Policy, 7
Principle of optimality
  theorem for adaptive control processes,
    43
  theorem for control processes, 13
  use in obtaining analytic result, 96

Queuing problem, 39, 71

Rocket example, 7
  problem on, 9
Roulette wheel example, 30

Search problem, 32
Sequential analysis, 4, 68, 89
Set of decision times, 6
Slot machine problem, 69
Solution
  adaptive control process problem, 30
  control process problem, 8
State, 6

State space
  conditions on, 8
  trajectory satisfies, 8
Statistical communication theory, 88
Stochastic process, 47
Stock disposal problem, 32, 70
Strategy, 27
Strategies
  set of, 29
Supervised learning, 73
  adaptive control process model for, 76
  as part of statistical communication
    theory, 88
  convergence in, 75
  generalizations of problem, 83
  model, 4
  numerical example, 82
Sufficient statistics, 78

Trajectory, 7
  a stochastic process, 28
Transducer, 72
  testing for optimum selection, 85
Transinformation, 84
Traveling salesman problem, 22
Two-armed bandit
  elementary problem, 49
  model, 3
  operations research model, 4
Two-armed bandit problem
  general problem, 65
  multi-armed and discounted, 59
  program and readout, 99, 105, 115
  numerical example, 56, 61
  solution to infinite play, 62
    algorithm, 65